DATA MINING

DATA MINING

PREDICTING PHISHING WEBSITES

M A Alhusain

KNOWLEDGE
BAKERS

KNOWLEDGE BAKERS

Cataloging in Publication Data--DK

Courtesy: D.K. Agencies (P) Ltd. <docinfo@dkagencies.com>

Alhusain, M. A., author.

Data mining : predicting phishing websites / by MA Alhusain.

pages cm

Includes bibliographical references.

ISBN 978-93-49964-94-5

1. Data mining. 2. Phishing--Prevention. I. Title.

LCC QA76.9.D343A44 2023 | DDC 006.312 23

ISBN: 978-93-49964-94-5

Published by : **Knowledge Bakers**
Pune, India
Email: knowledgebakers2019@gmail.com

ACKNOWLEDGMENTS

First and foremost, I thank God for granting me the strength, patience, good health, and ability to pursue this goal in my life.

I would like to express my deep thanks and gratitude to my supervisors Prof. Francisco Chiclana and Prof. Fadi Thabtah for their continuous support and everlasting encouragement during the long journey of my PhD study in the United Kingdom. My thesis would not have been possible without the guidance and the assistance of my supervisors who contributed to the completion of this study.

My sincere thanks go to Dr. Jenny Carter for her support in allowing me to further enhance my academic experiences through teaching modules and supervising master students at De Montfort University.

My thanks should also be extended to my family throughout the past four years. Very special thanks to my wife Neda'a. Her patience, encouragement and support over the period of study in UK made things much easier than expected.

Many thanks also go to my children who endured my absence and have always been granting me their love and support.

Finally, I would like also to thank all my friends especially; Sadir, Muhannad, Mohammad, and Ayad and all of my colleagues in the Centre for Computational Intelligence who support me during my study.

CONTENTS

LIST OF FIGURES

LIST OF TABLES

LIST OF ALGORITHMS

Page

List of Acronyms

AC	Associative Classification
AOL	America Online
ANN	Artificial Neural Network
APWG	Anti-Phishing Work Group
ARL	Average Rule Length
CANTINA	Carnegie Mellon Anti-phishing and Network Analysis Tool
CBA	Classification Based on Associations
CFS	Correlation Feature Set
DM	Data mining
DROP5	Decremental Reduction Optimization Procedure 5
eDRI	Enhanced Dynamic Rule Induction
ENN	Edited Nearest Neighbor
FFNN	Feed Forward NN
FOIL	First Order Inductive Learner
IG	Information Gain
IREP	Incremental Reduced Error Pruning
LERS	Learning from Examples Based on Rough Sets
MaT	Monitoring and Takedown
MDL	Minimum Description Length
NN	Neural Networks
OPUS	Optimized Pruning for Unordered Search
PILFER	Phishing Identification by Learning on Features of Email Received
PMCRI	Parallel Modular Classification Rule Induction
REP	Reduced Error Pruning
RIONA	Rule Induction with Optimal Neighbourhood Algorithm
RIDOR	Ripple Down Rule learner

SLIPPER	Simple Learner with Iterative Pruning to Produce Error Reduction
SSL	Secure Socket Layer
TF	Term Frequency
TF-IDF	Term Frequency Inverse Document Frequency
UCI	University of California Irvine
VV	Variable Value
RIPPER	Repeated Incremental Pruning to Produce Error Reduction
SVM	Support Vector Machine
WARL	Weighted Average Rule Length
WEKA	Waikato Environment for Knowledge Analysis
WOT	Web of Trust

Chapter 1

Introduction

1.1 Introduction

Data mining (DM), which is based on computing and mathematical sciences, is a common intelligence tool currently used by managers to perform key business decisions (Abdelhamid & Thabtah, 2014). Traditionally, data analysts used to spend a long time gathering data from multiple sources and little time was spent on the analysis due to limited computing resources. However, since the rapid development of computer networks and the hardware industry, analysts are spending more time on examining data, seeking useful concealed information. In fact, after the recent development of cloud computing, data collection, noise removal, data size, and data location are no longer obstacles. Data analysis, or DM, is concerned about finding hidden patterns from datasets that are useful for users, particularly managers, to conduct suitable planning (Thabtah & Hamoud, 2014).

One of the known DM tasks, which involves forecasting class labels in previously unseen data based on models (classifiers) learnt from training datasets, is classification (Hall, et al, 2009). Generally, classification is performed in two steps: constructing a model, which is often named the classifier, from a training dataset, and utilising the classifier to predict the value of the class of test data accurately. This type of learning is called supervised learning because while building of the classifier the learning is guided toward the class label. Common applications for classification are medical diagnoses, web security, and stock market analysis (Rameshkumar, et al, 2013; Mohammad, et al, 2015; Dash & Dash, 2016), etc.

One of the least studied approaches in classification is that which is based on separate and conquer learning (Pagallo & Haussler, 1990). This learning approach normally splits the training dataset into subsets based on classes and discovers the rules for each class through relationships between the input dataset's variables and class values. Once the complete set of rules is discovered, a classifier is formed by integrating the rules.

This thesis focuses on rule-based classification models, particularly

Covering approaches in DM. It proposes new dynamic Covering classification techniques that generate predictive models. These models contain knowledge represented as "if-then" rules, therefore they can be easily understood and maintained by decision makers. This chapter discusses the research problems under investigation, as well as the major contributions achieved. Lastly, we outline the structure of this PhD report.

1.2 Motivation

Supervised learning is one of the common learning approaches in machine learning (ML) and DM. In supervised learning, a model is normally constructed from a labelled set of examples called the training dataset. A typical training dataset consists of a set of variables (features) and a special variable called the class. For example, in medical diagnoses the class is the type of illness, and for credit card scoring the class is whether an applicant is granted credit or not. The primary purpose of the classifier is to accurately predict the class of an unseen dataset called the test dataset. There have been many different classification approaches proposed by researchers in the last few decades, including Decision Trees, Neural Network (NN), Support Vector Machine (SVM), Associative Classification (AC), and Covering among others (Quinlan, 1979; Mohammad, et al, 2013; Cortes & Vapnik, 1995; Thabtah, et al, 2004; Holt, 1993). The latter two approaches, AC and Covering, extract classifiers that consist of rules, which explain their wide applicability. However, in practise there are differences between AC and Covering, in particular in the way rules are induced and pruned. This thesis falls under the umbrella of Covering rule-based classification research.

There are a number of distinguishing features for Covering approaches in classification. The most dominant being that they are easy to understand models. Unlike models devised by traditional classification models such as NN, decision trees, or SVMs which normally require a domain expert to interpret them. The simplicity of Covering based models can be attributed to the straightforward rule format used, which novice users can easily interpret and utilise in decision making (Abdelhamid, 2015). Another noticeable feature of rule-based models is that users can easily maintain the classifier by adding, removing, or even amending rules when needed (Witten & Frank, 2005). This feature is hardly accomplished in other predictive models such as Decision Trees. Indeed, in decision trees the entire tree (classifier) must be reshaped if one tries to add a branch or partial tree to the existing tree, which is a complex process. Lastly, in Covering approaches, rules are ranked in a way that enables end users differentiate between highly effectives rules (mostly ranked at the top of the models) and less effective rules. This ranking is vital in choosing the rule or the set of rules that are able to predict the test cases, and therefore ranking influences the number of test data misclassifications.

The methodology in most existing Covering classification algorithms requires splitting the training dataset into a number of subsets based on class labels before the training phase begins. This approach was criticised by scholars in ML and DM because it builds local classifiers that belong to a subset of the training dataset rather than a complete training dataset (Abdelhamid & Thabtah, 2014). This criticism is attributed to the fact that rules are generated per class rather than from the entire classes at once, and in some cases a variable value belonging to multiple classes is considered only with the most frequent class during the training phase. This can also cause the generation of only single rules (one class per rule).

Another obvious problem associated with Covering techniques is the large numbers of rules induced, which normally leads to large classifiers (Witten & Frank, 2005). This problem appears because of the way the algorithm induces the rules; it keeps adding items to the rule's body until it covers only positive training data examples. In other words, the Covering algorithm typically produces many rules, each covering only a few examples of positive data, rather than rules covering large numbers of positive and negative data samples. The outcome of the aforementioned excessive learning is a classifier with several specific low data coverage rules that may overfit the training dataset. This may limit the use of this learning approach to domains that require a concise set of rules for decision making. Often managers in application domains such as medical diagnosis or credit card scoring prefer a summarised set of rules that they can manage instead of a high maintenance set of rules. There should be compromise between the number of rules and the predictive accuracy.

A more crucial issue that is worth investigating in Covering classification is the way rules are induced. When a rule is discovered, the algorithms directly delete its corresponding positive training data. The Covering algorithm then scans the adjusted training data to discover the next rule, produce it, remove its linked data, and then scan again the newly adjusted training data and so on (Abdelhamid & Thabtah, 2014). This repetitive process, especially the multiple data scans, consumes large computing resources and should be dynamically implemented.

This thesis investigates shortcomings associated with Covering algorithms in data mining. Specifically, we look at research issues including but not limited to the following:

1) Minimising computing resources, especially the number of passes over data examples, by proposing a dynamic learning technique.

2) Reducing the search space for both items and rules to make the training phase more efficient.
3) Generating smaller predictive models without hindering predictive power so end users can manage the classifiers in a straightforward manner.
4) Derive good quality rules, not necessarily with 100% accuracy, to reduce over-fitting and increase data coverage per rule.
5) Generating web security models based on Covering approach to serve as an anti-phishing rules set. This is since novice users usually prefer easy to understand rule to make them better understand phishing activities and features.
6) Identifying small, effective, phishing features.

This thesis involves only binary and multi-class classification problems since they are more common in classification applications. These problems produce datasets with examples, which are linked with one class. In other word, each instance in the training dataset is connected with one class value. The multi-label classification case is out of the scope of this thesis since they have different data format and requires special transformation. The scope is limited to the Covering approach in data mining, so other approaches related to generating rules such as Associative Classification are not explored. Associative Classification methods are based on association rule so they require the complete rule sets to be derived using unsupervised learning search before building the classifiers.

1.3 Covering Induction Process and Related Definitions

In Covering Classification approach, knowledge is discovered readily as the algorithm typically discovers rules per class label. For each class, i.e. C_1, the algorithm makes an empty rule (if empty then C_1) and looks for the best variable value (VV), such as Max VV_i, which leads to the largest gain according to a certain metric, or accuracy, with C_1. The rule's accuracy can be achieved by dividing the combined frequency of VV_i and the class, C_1, by the number of data examples that belong to VV_i in the training dataset.

The algorithm merges the VV_i with the largest accuracy with the rule's antecedent (left hand side), and proceeds to search for the next VV_{i+1} in the training dataset with which to append. The algorithm terminates building the rule when no more VV_i with acceptable accuracy can be found, and at that point the rule is generated and all of its positive examples that belong to C_1 are discarded. The algorithm continues discovering the rules from the remaining data examples that belong to class C_1 until that set becomes empty, or no VV_i when integrated with C_1 has an acceptable accuracy. When this occurs, the algorithm moves on to the next class data set and repeats the aforementioned steps until the complete rules set is derived.

The problem can be formally defined as follows: Given a training dataset T, which has x distinct columns (variables) $V_1, V_2, \ldots, V_n$, one of which is the class, i.e. *class*. The size of T is $|T|$. A variable in T may be nominal, which means it takes a value from a predefined set of values, or continuous. Nominal attribute values are mapped to a set of positive integers whereas continuous attributes are pre-processed by discretising their values using any discretisation method.

The aim is to construct a classifier from T, e.g. *Classfiier* : $V \rightarrow class$, which forecasts the class of a previously unseen dataset.

The classification method investigated employs a user threshold called ***freq***. This threshold serves as a fine line to distinguish strong rule items <item, class> from weak ones based on their computed occurrences in T. Any rule item whose frequency bigger than the *freq* is called a strong rule item, otherwise being referred to as a weak rule item. Below are the main concepts used throughout the thesis and their definitions.

Definition 1: A variable plus its denoted values name (V_i, v_i) is known as *item.*

Definition 2: A *training example* in T is a row consisting of variable values $(V_{j1}, v_{j1}), \ldots, (V_{jv}, v_{jv})$, plus a class denoted by $class_j$.

Definition 3: A *rule item r* has the format $|body, c|$, where *body* is a set of disjointed items and *c* is a class value.

Definition 4: The frequency threshold (*freq*) is a predefined threshold given by the end user.

Definition 5: The body frequency $|body\_Freq|$ of a *rule item r in T* is the number of training examples in T that matches *r's body.*

Definition 6: The frequency of a *rule item r in T* (*ruleitem_freq*) is the number of training examples in T that matches *r*.

Definition 7: A *rule item r* passes the *freq* threshold if, *r*'s $|body\_Freq| / |T| \geq freq$. Such a *rule item* is said to be a strong *rule item.*

Definition 8: The rule strength |rule_strength| is a predefined threshold given by the end user.

Definition 9: A rule *r* has a strength which is defined as $|ruleitem\_freq| / |body\_Freq|$.

Definition 10: A rule *r* can be generated when *r*'s $|strength| >= |rule\_srtength|$.

Definition 11: A rule takes the form: *r*: *body* →*class*, where body is a set of disjoint variable values and the consequent is a class value. The format of the rule is: $v_1 \wedge v_2 \wedge \ldots \wedge v_n \rightarrow class_1$

1.4 Thesis Research Issues

Different issues related to rule based classification approaches, particularly Covering classification, are discussed in this research. These are:

Generic Issues

- Developing a dynamic rule discovery process.
- Improving the performance of Covering techniques.
  - The reduction of the search space of items during training phase.
  - Generating a concise set of rules by removing redundant rules in the classifier.
  - Minimising over-fitting the training dataset.
  - Improving training time required to construct models.

Domain Specific Issues

- Covering classifiers as an anti-phishing tool.
- Determining the most effective features related to phishing.

1.4.1 Generic Issues

GI-Issue 1: The Methodology to Discover the Rules

In most rule-based classification the process of rule discovery is accomplished in a separate phase after the algorithm goes over the training dataset to record item and class occurrences. For each available class, these algorithms add the best item to the class according to a certain measure of the rule's body. When the rule reaches the desirable accuracy, then it gets derived and these algorithms discard all classified training data examples that belong to the rule. The algorithm then proceeds to discover other potential rules for the class at hand from remaining data examples by reviewing the complete dataset again, wasting runtime and memory. This step is necessary, however, since the frequency computed earlier for the items and the class labels for some potential rules have changed. This process not only is resource demanding, but can also cause the algorithm to crash when the data is highly dimensional. A simple experiment using the PRISM Covering algorithm implemented in WEKA ML tool against a "Segment Challenge" dataset downloaded from the University of Irvine data repository was conducted to record how many data examples this algorithm must go over to learn the rules (Cendrowska, 1987; Hall, et al., 2009; Licham, 2013). The "Segment Challenge" dataset consisted of twenty variables and 1500 examples. It turns out that this methodology must scan 7,147,542 data examples to find the rules. This number can be of higher magnitude if datasets with larger dimensionalities are used, which is a burden in big data domains.

Contribution

There should be a mechanism embedded in the training phase which ensures that the rank and frequency of potential rules are amended in a dynamic manner without any repetitive training data scans. This will make the process of rule induction efficient, not static every time a rule is derived. Therefore, whenever a rule is inserted into the classifier all remaining potential rule <item,class> occurrences should be updated extemporaneously based on data removal. We developed a new dynamic rule learning that substantially reduces the number of examples scanned. Our method, which we called Enhanced Dynamic Rule Induction (eDRI) Qabajeh, et al, (2014) is presented in Chapter three.

Experimentations using different classification datasets that belong to applications beyond medicine, credit card scoring, finance, agriculture, politics, and others from UCI data repository (Lichman, 2013) have been conducted to evaluate the proposed eDRI. Chapters four and five give further detail on the compared algorithms, experiment settings, evaluation measures, and data characteristics. The results show that eDRI outperforms other classic ML algorithms as well as rule-based algorithms in regard to the considered evaluation measures. Moreover, models produced by eDRI not only have good predictive power, but they are also efficient during data processing and contain small effective rules for decision making. These rules have low to no redundancy since they do not overlap in the training examples and normally require a lower number of data scans to be found that the compared Covering algorithms.

GI-Issue 2: Enhancing the Performance of the Classification Models (Classifiers)

We investigated different problems in rule-based classification related primarily to the process of rule discovery and classifier content. These issues are:

- Reducing the search space for items
- Decreasing the number of rules in the classifier so users can better understand and use them

- Improving test data prediction by using multiple rules based on weights

A. *Cutting down the items search space during the training phase*

As mentioned before, in typical rule-based classifiers such as Covering the methodology to find the rule necessitates that after generating each rule a full scan is conducted on the remaining data examples. This is necessary to record item frequencies after data removal; hence, the search space of items is massive and consequently consumes large computing resources. In fact, the number of items that can participate in rule induction should be minimised since many of these items have no statistical significance and removing them will improve the process of creating rules. Thus, there is a need for a mechanism that discriminates between items to reduce the numbers of items in the search space during the training phase.

Contribution

The rule discovery process proposed in the enhanced Dynamic Rule Induction algorithm (eDRI) minimised the number of items that can be offered to the rules. We have introduced a frequency threshold (freq, see Definition 6) that discriminates among items based on their occurrence with the item and class in the training dataset. The freq threshold serves as an indicator that distinguishes weak items from strong items. Any item plus class frequency must pass the freq threshold so it can be considered by the rules. This has reduced the search space for items during rule building by discarding weak items early on and in a dynamic manner. In fact, strong items can become weak whenever a rule is derived if they appeared in the discarded rule's data. Experimental studies in Qabajeh et al, (2015) showed an improvement in use of computing resources, especially run time reduction and number of passes required for building the classifier. These results are documented in Chapter 4.

B. *Reducing the number of rules for better controllability*

One of the drawbacks associated with Covering algorithms is that despite being simple classifiers, the number of rules they produce is large. This can be attributed to the way the algorithm discovers rules in that each correlation between class and variable is investigated and which leads to the production of several specific rules (covering very limited data examples). Most Covering algorithms, like PRISM and its successors, keep appending variables into the rules body until the rule achieves a zero error rate. This can result in just covering one or two training data examples, which limits generalising rule performance across wider data applications.

To overcome this issue, we investigated this problem and proposed a rule pruning procedure to cut down the number of potential rules and reduce the classifier size. It should be noted that search induction learning algorithms such as PART, IREP and RIPPER use extensive such as backward and minimum description length principle pruning to cut down the size of the classifiers (Frank & Witten, 1998). However, these algorithms originally employ a divide and conquer approach to learning trees that then are converted into classification rules. Thus, it is impractical and incorrect to consider them Covering approaches despite generating rules.

Contribution

We have developed a pruning method that ensures each potential rule has a true rank during the process of rule generation. When a rule is constructed, and its data examples are discarded, we immediately adjust the potential rule ranks which help us to determine weak rules early on and remove them. The resulting analysis against a large data collection from the UCI repository showed moderate size models being consistently generated by our dynamic technique when compared with those generated from decision trees and other Covering algorithms. Interestingly, the models produced by eDRI had a lower number of rules yet the predictive performance was sustained for the majority of datasets. Furthermore, in phishing detection, eDRI was able to construct small set of useful anti-phishing rules that contained new and interesting information for security experts and online users. Lastly, when the phishing data was processed, the accuracy of our incremental model was only slightly affected, which shows constant predictive performance of eDRI on different domains data. All the aforementioned results are elaborated in details in Chapters 4 and 5.

C. Avoiding Over-fitting and Under-fitting

One of the important issue in supervised learning predictive models is over-fitting the training dataset. This problem is obvious in Covering approach when the learning algorithm excessively trains against the input dataset particularly when the algorithm allows the generation of only perfect rules, which is true case with PRISM. These rules usually are associated with maximum accuracy against the training dataset, yet they are not general rules that can be widely applied on an unseen dataset. Therefore, the predictive performance of the rules when they are applied on a test dataset is questionable. Another obvious example is the REP algorithm that generates a complete rule set and then invokes an evaluation step to cut down redundant rules (Fürnkranz & Widmer, 1994). There should be a mechanism for learning to be stopped so that training datasets will not be overfitted. This mechanism guarantees realistic rules in the final classifier that are linked with insufficient data representation being discarded early during the training phase.

Contribution

We posit that generating high quality rules, not necessarily perfect ones but rather those with larger classified training examples, is advantageous. This can be accomplished using user statistical measure such as rule strength (see definition 9). The rule strength separates acceptable from unacceptable rules based on both data coverage and items' frequency within the class label. This is important because applications such as phishing classification necessitates small yet highly correlated rules in the classifiers. Rule strength serves as a point at which we stop adding items into the rule's body and therefore reduce over-fitting the training data by allowing for a slight error per rule while maintaining larger data coverage. Moreover, rule frequency dynamically during the training phase contributes to reducing over-fitting. This is since items within potential rules that fail to accomplish the frequency threshold will be discarded, and therefore these rules will not be considered during the training phase and consequently further shrinks the search space.

1.4.2 Domain Specific Issues

Issue 3: Rules Set as Effective Anti-Phishing Tool

Phishing is one of the most serious online frauds, costing different stakeholders including banks, online users, governments, and other organisations severe financial damages. This fraud often entails designing and implementing a replica website that maintains visual similarity to that of an existing trusted organisation (Aburrous, et al, 2010). The aim of phishing is to unlawfully obtain user login data in order to access their financial information (Mohammad, et al, 2015). One promising research direction to minimise the risks of phishing is using predictive ML and DM algorithms. Despite promising results with reference to classification accuracy of the aforementioned approaches, the models they produce are usually complicated in a way that novice users might find difficult to interpret and use them for making decisions. Moreover, adding to or amending the model requires careful implementation and is often a complex process that end users might be unable to perform (Abdelhamid, et al, 2014).

Contribution

A model with straightforward knowledge represented in simple rules is obviously advantageous for end users. Not only the model could be used to detect phishing activities, but it could also serve as a decision making tool in which the user can readily understand the relationship between phishing features and the website's class, i.e. phishy or legitimate. There has been recent research on the use of rule-based classifiers utilising associative classification reported in Abdelhamid (2015) and Abdelhamid, et al (2014). However, these approaches suffer from noticeable drawbacks, which are not limited to just the size of the classifiers derived but also the inherited problems from association rules such rule redundancy. A Covering approach that enhances the process of rule discovery and classifier building as discussed in Section 1.4.1 and removes data overlapping can be an effective anti-phishing tool. To date, there is very limited work on anti-phishing using covering classification.

A case study, Chapter 5, that investigates phishing as a classification problem is reported. This chapter contains critical analysis of phishing approaches, from traditional to education and intelligent ones, with their associated benefits, disadvantages and ways forward in combatting phishing. More importantly, a comprehensive experimental study on real phishing websites is reported. eDRI's performance as a phishing detection technique is compared against the performance of other ML techniques to reveal the strengths and weaknesses of eDRI.

1.5 Thesis Structure

This thesis consists of five chapters including the present one. Chapter two critically analyses Covering techniques in data mining, separate and conquer, divide and conquer, and their mutually common algorithms. We have included induction learning approaches as well. Chapter three is devoted to the proposed Covering algorithm (eDRI) for constructing rule-based predictive models. In chapter three, we present the dynamic process mechanism, pruning procedure, class forecasting method and a detailed example that explains eDRI insight. Chapter four contains the implementation of our newly enhanced dynamic method during the training phase. We also present the evaluation measures used and phase details of our algorithm alongside a complete example. In addition, the results on the UCI datasets and their analysis are discussed in chapter four. These are not only limited to large variations of UCI datasets, large selection of rule and non-rule based ML algorithms, and evaluation measure results analysis. Real data experimentations related to phishing and using a number of ML techniques have been conducted in chapter five. This section also reviews and critically analyses the phishing problem and traditional solutions based on legal, educational, and intelligent anti-phishing tools. A large experimental analysis is then conducted to evaluate the benefits and disadvantages of the proposed algorithm. Lastly, chapter six concludes the thesis by responding to the research issues raised in chapter one and highlights the main contributions of this thesis, further pinpointing possible future research paths.

Chapter 2

Covering and Induction Approaches in Classification

2.1 Introduction

One of the primary tasks in ML is predictive classification (Diebold and Mariano, 1995; Abdelhamid, et al., 2016). In classification, a predictive model called the categoriser/classifier is typically constructed from a training dataset. This model is then applied to test data in order to predict the class as accurately as possible. Effectiveness of the model is measured mainly on its performance in predicting test cases (Abdelhamid and Thabtah, 2014). The less number of misclassifications the model scores on test data cases the higher predictive power the model possesses and therefore its performance can be generalised.

There are two forms of classification problems; single label and multi-label (Fürnkranz, et al., 2008). The former can also be classified into two types: binary and multi-class classification. In binary classification, training data can only be linked to a single class value from two possible values in the training dataset, while in multi-class classification each training instance is linked with one class value from two or more possible values. On the other hand, multi-label classification problems necessitate that each training instance has multiple class values, and thus this type of data requires careful transformation (Taiwiah and Sheng, 2013). This is since each instance must be repeated with each class label. More details on the different classification problem types are given in Section 2.2.

Different intelligent learning methodologies based on DM and ML have been developed and disseminated to deal with the classification problem. Among these we have support vector machine (SVM), neural network (NN), associative classification (AC), boosting, regression trees, probabilistic, decision trees, induction learning, and Covering among others (Joachims, 1999; Mohammad, et al, 2012; Liu, et al, 1998; Abdelhamid, 2015; Freund & Schapire, 1997; Breiman, et al, 1984; Duda, 1973; Quinlan, 2002; Cohen, 1995; Bramer, 2002). The majority of the learning approaches solve classification by decomposing the problem into two phases discussed berfore

- Learning Phase: This phase involves learning a model from the training dataset.
- Prediction Phase: Using the learned model to predict the class label value in a test dataset.

In this thesis we deal with Covering classification, yet in Section 2.3 other common induction ML approaches that are suitable to solve the classification problem are still discussed.

As mentioned in Chapter One, Covering classification is one of the most important learning approaches that generate predictive models with "If-Then" rules. These "If-Then" rules can be utilised as a knowledge base by end-users for decision making (Qabajeh, et al., 2015). A typical Covering algorithm learns rules one by one and for each available class in the training dataset. The algorithm initially creates an empty rule for a specific class [If empty, then Class 1]. Then using certain mathematical metrics such as accuracy, foil gain, etc., append one item to the empty rule body (left hand side) (Cendrowska, 1987; Quinlan, 1990). The algorithm keeps appending items until the rule metric cannot be further enhanced, at which point the rule is inserted into the classifier. In the same manner, the algorithm continues deriving the rules for the classes until a stopping condition is met.

Three noticeable characteristics exist of a Covering approach in solving classification problems (Clark & Boswell, 1991; Domingos, 1996a; Grzymala-Busse, 2010; Abdelhamid & Thabtah, 2014)

- The learned model contains simple yet highly influential rules for decision making that end-users can easily comprehend and manage.
- Adding new knowledge (rules) can be done in a straightforward manner.
- The way models are constructed is based on easy searches.

This chapter defines classification problems, followed by discussing classic induction learning approaches. More importantly, the focus will be on Covering classification methods as well as other common rule-based classifiers. The chapter structure is as follows: the classification problem and its main steps are discussed in Section 2.2; classic learning strategies dealing with predictive tasks such as classification are briefly discussed in Section 2.3, while section 2.4 critically analyses Covering classification approaches along with other common rule-based classifiers in DM and ML. Moreover, we also introduce non-rule classification methods such as Boosting, probabilistic, SVM, AC and NN in section 2.5. Finally, conclusions are drawn in section 2.6.

2.2 The Classification Problem and Its Variations

In Chapter One, we have discussed the main terminologies related to the classification problem, specifically those connected with the Covering approach. In this section, we define single label classification problems (both binary and multi-class) in the ML context. Figure 2.1 depicts the essential steps that any learning algorithm goes through in order to construct models and apply them to predicting test cases. Input will consist of a set of labelled variables, including the class, called the training dataset. A learning algorithm then processes the training dataset to derive the classifier. This classifier is then applied to a test dataset to measure its prediction power (see Figure 2.1).

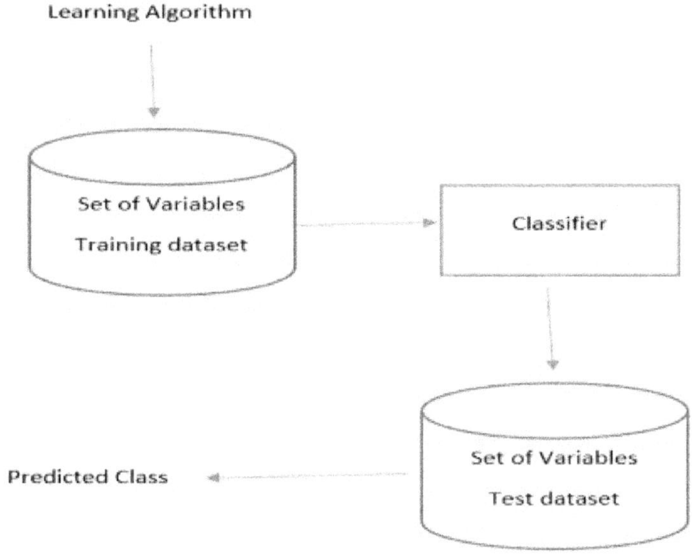

Figure 2.1 Classification steps

Multi-label and multi-output classifications are variants of the classification problem where multiple target labels must be assigned to each test case (Veloso, et al, 2007). Multi-label classification should not be confused with multi-class classification, which is the problem of categorising instances into one of more than two classes. Formally, multi-label learning can be phrased as the problem of finding a model that maps inputs (x) to binary vectors (y), rather than scalar outputs as in the ordinary classification problem (Tsoumakas, et al, 2010).

In binary classification and according to Bradley, et al (1998), given an input dataset taking the format of (X_i, C_i) we can assign the given vector $x_i \in \Re^n$ into one of two disjoint sets A or C in a dimensional variable space where n. $X_i \in \Re^n$ is the ith data instance and $C_i \in \{1, \ldots.., K)$ is the ith class. The ultimate goal is to derive a function, F, that maximises the chance that $F(x) = C_i$ for an unseen data instance. The classification problem can basically be transformed into a two class problem in which the class c_i can be either 0 or 1. This simple format of the problem is called binary classification. Binary classification function, F(x), has the following format:

$$F(x) = \begin{Bmatrix} 0 & x \in A \\ 1 & x \in C \end{Bmatrix}$$

Multiclass classification deals with predictive problems that involve more than two values for the class label.This is because the training dataset in multi-label classification, such as an image or a science, contains instances where each may be associated with multiple labels rather than a single class as in multi-class or binary classification (Tsoumakas, et al, 2010). Tables 2.1a-2.1b depict multi-label and multi-class datasets, respectively. Table 2.1b is the transformation of Table 2.1a from multi to single label classification.

Often multi-class and multi-label datasets require data transformation, so strategies that transform their datasets into binary classification are essential. Two of the common data conversion strategies from multi-class and multi-label into a binary classification are one-versus-one and one-versus-all (Fürnkranz, et al, 2008). The latter essentially learns single classifiers for each class label. While learning the classifier for a class C, all training instances that belong to C are considered positive instances and the remaining are negative instances. Thus, when a test case is about to be predicted, a score based on the required single classifiers is calculated to come up with the class decision (Thabtah et al, 2006). One shortcoming associated with the one-versus-all approach, however, is when the training dataset is imbalanced because it potentially creates a computed score that can be arisen with biased to the majority class. The other data conversion strategy, the one-versus-one approach, necessitates producing $N(N-1)/2$ binary models for the N multi-class problem (dataset)(Thabtah, et al, 2006). When test data is about to be predicted, a voting mechanism is employed in which all $N(N-1)/2$ are utilised and the class with the largest correct predictions is assigned by the multiple binary models.

According to Tsoumakas, et al, (2010), for multi-label classification problems two data processing methods are also used; problem transformation and algorithm adaptation. The problem transformation consists of converting of the available multi-label dataset into binary classification sets in that each set can then be dealt with using any binary classification methods. In other words, the problem transformation strategy can be seen as a pre-processing phase which ensures that each training instance in the original multi-label dataset is transformed into multiple single label instances. On the other hand, algorithm adaptation strategy employs a specific algorithm to learn multi-label models directly from the multi-label classification datasets without the use of data pre-processing phase. This approach to learning from multi-label problems is domain specific, and requires different learning strategies to acknowledge the fact that an instance has multiple class labels and the prediction involves assigning multiple labels to the test data.

Table 2.1A Multi-label dataset

Variable 1	Variable 2	Variable 3	Class
X1	Y1	Z2	C1, C3
X1	Y1	Z2	C1, C3
X2	Y1	Z3	C1,C2, C4
X2	Y2	Z3	C2
X1	Y3	Z2	C3,C4
X2	Y2	Z1	C2

Table 2.1B Multi-class dataset transformed from Table 2.1A

Variable 1	Variable 2	Variable 3	Class
X1	Y1	Z2	C1
X1	Y1	Z2	C3
X1	Y1	Z2	C1
X1	Y1	Z2	C3
X2	Y1	Z3	C1
X2	Y1	Z3	C2
X2	Y1	Z3	C4
X2	Y2	Z3	C2
X1	Y3	Z2	C3
X1	Y3	Z2	C4
X2	Y2	Z1	C2

This thesis focuses only on binary and multi-class classification, and leaves multi-label classification out of scope of this study since this requires effort for additional PhD.

2.3 Learning Strategies in Classification

2.3.1 Divide and Conquer

This classification approach splits data into subsets, which are further divided into smaller divisions until the algorithm believes that the subsets are adequately homogenous or the stopping criterion has been satisfied (Quinlan, 1979). To illustrate this, imagine a root node being the entire dataset. Subsequently, the algorithm specifies a variable in the dataset to start the first split. In an ideal situation the algorithm starts with the feature that maximises discrimination among the outcome classes in the dataset. Decision tree algorithms which apply divide and conquer use Gini index or Information Gain metrics to select the variable that can be the root node (Zaremotlagh, S., & Hezarkhani, 2017; Gini, 1999; Quinlan, 1986). The cases are next grouped into classes based on the splitting of the variable. This process is repeated and the algorithm continues to separate (divide) and group cases while excluding heterogeneous data cases (conquer).

Decision tree algorithms are common divide and conquer methods that classify cases based on their attributes' values. Nodes represent variables to be classified, and branches represent the values that variables take (see Figure 2.2). Cases are classified beginning at the root node. Figure 2.2 represents a decision tree for a tax evasion classification problem (Tan, et al, 2005). Based on a number of attributes (tax refund and taxable income along with demographic information and marital status), the user wants to predict whether citizens cheated on their tax forms.

Note that the attribute that best classifies the cases is represented by the root node feature. In this example, marital status seems to be the best one. Married people did not cheat on their taxes as the first node of the tree indicates. The process of classification based on best features continues until a parsimonious representation is obtained. Decision trees are useful given their representations for classification problems, which constitute a large portion of everyday applications. The detailed process of how to construct decision trees from training dataset and common methods used are given below.

Imagine that you are a university administrator working for a public university. Your role is to decide whether new programmes should be approved for enrolment in the next academic year. Upon your return from holiday, your work station is full of requests from various colleges. Realising that you do not possess sufficient time, you decide to construct a decision tree to help make the decision of categorising the requests into three groups: Success, Revision and Failure. To construct the decision tree, you explore factors associated with the 20 most successful and 20 worst programmes at the university during the past 25 years. You realise that there is a strong correlation between the programmes' budget, size, and college affiliation of programme with the probability of its success or failure. Beginning with the entire dataset, the root node, we can select programme budget as the first feature by which to split programmes. Programmes can be classified as having a cost of more than $1 million or less. Second, among the programmes with a cost of less than $1 million, we can classify them based on their college affiliation. We could further split the programmes, yet over-fitting a tree is non-advisable. Over-fitting occurs when someone is trying to over learn by building larger trees in order to maximise the predictive performance of the model (tree) on the training dataset rather than on the test dataset (Hall, et al, 2009). You could therefore decide that programmes meeting a cost of less than $1 million dollars within a technical college affiliation are more successful, and thus approve them.

Despite its applicability, the divide and conquer approach, it suffers from a number of limitations (Kothari & Dong, 2000; Leung, 2007; Grzymala-Busse, 2010). First, it tends to split the data based on variables with a large number of levels. Second, it is easy to over-fit or under-fit a model for classifying instances in a given dataset. Under-fitting the model occurs when someone stops building the tree early, and thus produces a model with limited performance (Witten & Frank, 2005; Jankowski & Jackowski, 2014). Third, the divide and conquer approach is problematic when it comes to modelling complex relationships since it relies on axis-parallel splits. Moreover, the output produced, especially in large trees, can be difficult to interpret and understand by non-technical audiences. In this event, the user will not be able to manage the tree and therefore will not be able to use it for decision making. Finally, a key weakness of the divide-and-conquer approach is its sensitivity to the input data (Leung, 2007; Grzymala-Busse, 2010). A small change in the input may result in significant changes in the decision trees produced. Introducing new cases into the dataset may result in redrawing the whole tree altogether. Despite the easy constructing decision trees, they tend to be more complex compared to other classification approaches, i.e. separate-and-conquer (see section 2.3.2) (Fürnkranz, 1999; Brookhouse & Otero, 2016), which This learning approach produces an easier representation of results, thereby making them superior to divide-and-conquer, particularly in their final models understanding.

2.3.2 ID3 Decision Algorithm

Decision trees in classification spread out after the dissemination of the ID3 algorithm (Quinlan, 1979). A decision tree can be constructed based on the available features inside the training dataset, primarily using the equation for Information Gain (IG) (Equation 2.1). The tree is built using the ID3 algorithm by initially selecting the attribute with the highest computed IG among all available attributes in the training dataset as a root node. IG is computed using Entropy (Equation 2.2) which denotes how informative a feature is in splitting the training instances according to the target attribute values. After choosing the root node, the algorithm computes IG repeatedly for the remaining attributes, excluding the root node until the tree cannot be divided any further or all training instances for an attribute belong to one class (Quinlan, 1993).

$$IG\ (T, f) = Entropy\ (T) - \sum((|T_f| \ / \ |T|) * Entropy\ T_f)) \qquad (2.1)$$

Where $$Entropy\ (T) = \sum -P_c \log_2 P_c \qquad (2.2)$$

where

P_v = Probability that T belongs to class l.

T_f = Subset of T for which feature F has value f_a

$|T_f|$ = Number of examples in T_f, and $|T|$ = Size of T.

A tree constructed by ID3 can be transformed into a rules set in which a path in that tree links the root node to any leaf making a rule. Figure 2.2 displays a tree that contains four rules that were constructed based on three features and a target attribute (class).

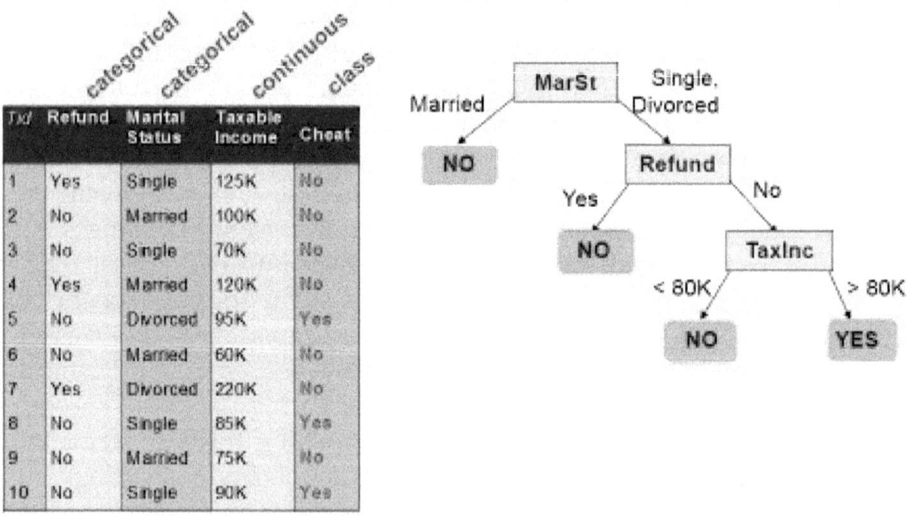

Tid	Refund	Marital Status	Taxable Income	Cheat
1	Yes	Single	125K	No
2	No	Married	100K	No
3	No	Single	70K	No
4	Yes	Married	120K	No
5	No	Divorced	95K	Yes
6	No	Married	60K	No
7	Yes	Divorced	220K	No
8	No	Single	85K	Yes
9	No	Married	75K	No
10	No	Single	90K	Yes

Figure. 2.2 Decision tree example (Tan, et al, 2005)

2.3.3 Separate and Conquer

Separate-and-Conquer is a learning strategy famous within the rule-based classification (Pagallo & Haussler, 1990; Fürnkranz, 1999; Rijnbeek & Kors, 2010). The classification problem is characterised by a number of positive and negative cases of a specific concept, normally called the class, and a set number of attributes inside a training dataset. The separate-and-conquer algorithm discovers the class in the form of explicit rules, or knowledge, composed of tests derived from values of attributes in the training dataset. The resulting set of rules should be able to explain instances of the target class and discriminate other instances that do not belong to it. Many approaches have been developed to solve this problem including covering and induction learning (Stahl & Bramer, 2014; Pazzani & Kibler, 1992).

Covering approaches are popular within rule-based classification. Despite their limited numbers of algorithms, they still constitute a viable approach in classification problems (Abdelhamid & Thabtah, 2014). Covering approaches are present in many free tools developed by the ML community such as R and WEKA (R Development Team, 2008; Hall, et al., 2009). Covering approaches have a number of advantages over other approaches of rule-based classification especially the simple to define outcome rules. However, before we delve into their strengths, we need to describe their basic strategy in classification problems.

Covering algorithms are similar in the logic they adapt when solving a classification problem. This family of rules learning share a common feature: a separate-and-conquer algorithm that searches for a rule that predicts part of the training set, while are isolated those, and then conquers the remaining cases learning new rules until no cases are left in the dataset. This approach guarantees that each case in the training set is covered by at least one rule (Grzymala-Busse, 2010). Pagallo & Haussler (1990) developed the term separate-and-conquer in reference to the common feature of those algorithms. Other authors have referred to this approach in different ways, "separate-and-conquer," "Covering," or "greedy" just to name few.

Covering approaches are characterised by a strong representation bias compared to other approaches such as classification tree algorithms (Fürnkranz, 1999). A simple rule is required to perform the same task a complicated tree would achieve. The separate and conquer approach tries to maximize the purity of a leaf when it attempts to generate a rule. On the other hand, classification trees look for the average purity when splitting a node. However, the rules do not possess the redundancy problem of the association rules approach. The underlying logic behind separate-and-conquer is to generate a parsimonious set of rules capable of classifying new cases by distinct rules within a dataset (Pagallo & Haussler, 1990). Covering approaches are more sensitive to collision problems when an instance could be classified into more than one rule (Witten & Frank, 2005).

Covering approaches work either in a bottom-up or top-down fashion (Grzymala-Busse, 2010). The bottom-up strategy begins with a positive case by attempting to generalise the rule through the removal of conditions. The basic assumption here is that the rule must cover the largest group of positive cases and the fewest number of negative cases. While valuable, this strategy is tedious and does not perform well in noisy datasets (Stahl and Bramer, 2012). The top-down strategy resembles the classification tree model. It aspires to maximise the purity of the leaf on a branch of a tree. Computationally, this strategy consumes equivalent time to that of the classification tree approach (Witten & Frank, 2005). Hereunder are the main advantages and disadvantages associated with traditional Covering approaches based on the comprehensive review we have conducted:

Advantages

1) The simplicity of rules generation in which only one metric is computed to decide rule significance (Franczak, 2002).
2) The classifier contains easy rules which can empower decision makers, particularly in domain applications that necessitate quick and easy interpretations (Abdelhamid, et al, 2014; Abdelhamid and Thabtah, 2014)
3) Straightforward implementation, especially with available computing resources (Stahl and Bramer, 2014).
4) The predictive power of its classifier can be seen as acceptable when contrasted to other classic DM and ML approaches such as search methods, decision trees, NN, Boosting, Regression, AC, and many others (Bramer, 2002; Thabtah et al, 2007).

Disadvantages

1) Noisy datasets that contain incomplete attributes and missing values. This can be seen as a major challenge since no clear mechanism on handling noise is presented (Grzymala-Busse, 2010). Currently, adopted approaches from information theory are used to handle noise as in Bramer (2000), Stahl and Bramer (2014), and Cohen (1995).
2) No clear rule pruning methodology is present. This may lead to the discovery of large numbers of rules and therefore combinatorial explosion (Abdelhamid & Thabtah, 2014). There is a high demand on pruning methods to cut down the number of rules without hinder-

ing overall performance of the classifiers (Abdehamid, et al, 2014). One possible promising solution is pruning methods adopted by induction learning algorithms such as PART, IREP, and RIPPER because they use reduce-error pruning methods and therefore decrease the number of rules.

3) Conflicting rules during prediction. There is no clear mechanism on how to resolve conflicting rules (Grzymala-Busse, 2010). Currently the choice is random and favours class labels with the largest frequency linked with the item rather than keep multiple class labels per item.

4) Breaking ties among item frequencies while building a rule. When two or more items having the same frequency, Covering algorithms such as PRISM look at their denominator in the rule accuracy formula. Yet, sometimes these items have similar accuracy and denominators, which makes the choice random. Arbitrary selection without scientific justification can result in biased decisions and may not be useful for overall algorithm performance (Thabtah, et al, 2007).

2.4 Common Covering and Induction Algorithms

There are many separate-and-conquer algorithms in the literature, the majority of which are local algorithms such as PRISM and a few are global algorithms like our algorithm eDRI (Chapter 3). In global Covering algorithms, all variable values are used in the search space together with the class variable. Whereas in local Covering algorithms the attribute-value pairs are used as the search space. In this section, we highlight common Covering algorithms and then shed light on other approaches that generate rule-based classifiers similar to Covering approaches.

2.4.1 PRISM Algorithm and its Successors

Covering methods such as PRISM derive rules one by one from the training dataset (Cendrowska, 1987). Normally these algorithms divide the training dataset into subsets representing each one of the target classes. The instances in one subset are considered positive to that subset's class label and negative for all other class labels. To make rules, the Covering algorithm is applied on each subset. A rule is made by attributing values to its body using a statistical measure such as rule's accuracy. A rule building stops upon reaching a certain condition. Once a rule is generated, all of the training data instances covered by it are removed and the algorithm discovers other rules in the same manner. Termination of the data processing is typically occurred when the training dataset has no remaining data instances or no rules with an acceptable accuracy can be found.

PRISM is a key algorithm for building classification models that contain simple yet easy to understand rules. This algorithm was developed in 1987, using data examples which are separated using the available class labels. Specifically, for each class (wi) data sample, an empty rule (If nothing then w1) is formed. The algorithm computes the frequency of each attribute value linked with that class, and appends the attribute value with the largest frequency to the current rule's body. PRISM terminates the process of building a rule when that rule has an accuracy of 100%. Meaning this algorithm favours very specific rules that have no negative instances. When the rule is generated, the algorithm continues processing the remaining possible rules from w1's data subset until it becomes empty or no additional attribute value can be found with an acceptable accuracy. At that point, PRISM moves to the second class data subset, repeating the same steps. When the training dataset is evaluated, the algorithm finally stops and the rules formed construct the classifier. Algorithm 2.1 below depicts the major steps of the PRISM algorithm. One of the major challenges faced by PRISM is noisy training datasets. These are datasets that contain missing values, data redundancy, and incomplete data examples. Bramer (2000) has developed a modified version of PRISM, called N-PRISM to handle noisy datasets as well as focusing on maximising prediction accuracy. N-PRISM has been implemented using Java within an R package. Experimental tests using eight datasets from the UCI have showed consistent performance of N-PRISM when compared to classic PRISM, particularly on classification accuracy obtained from noisy and noise-free datasets (Lichman, 2013).

It should be noted that the difference between the singular PRISM and N-PRISM is very minimal.

Stahl and Bramer (2012) developed another modified PRISM method based on a previously developed pre-pruning method called J-Pruning. Its purpose was to reduce over-fitting on the training dataset. J-Pruning is based on an information theory test performed typically in decision trees by measuring the significance of removing an attribute from the rule's body. This algorithm was developed to address rule pruning issues during the process of building up rules for each class label in PRISM (Stahl & Bramer, 2012). Experimental results using sixteen UCI datasets (Lichman, 2013) have shown decreases of the number of items per rule.

In 2015, Othman and Bryant (2015) investigated instance reduction methods as a paradigm for rule pruning in induction algorithms. They claimed that minimising the training dataset to the subset needed to learn the rules is one way to shrink the search space. The authors applied three common instance reduction methods, namely DROP5, ENN, and AllKnn, on a limited number of datasets to evaluate their impact on the resulting classifiers. Results obtained, although limited, can be seen as a promising direction of using instance reduction methods as a pre-pruning phase in induction learning.

Input: Training dataset T
Output: A classifier that consists of If-Then rules

1. for each subset Ti in T that belong to wi Do
2. make an empty rule, r_j: If Empty then wi
3. calculate Attx in Ti p(w_i = i | Attx);
4. append the Attx with the largest p(w_i = i | Attx) to the body of r_j
5. repeat steps 1.1-1.3 until r_j has 100% accuracy or no longer can be improved
6. generate r_j and insert it into the classifier
7. discard all data examples from Ti that match r_j's body
8. continue producing rules until Ti is empty or no rule with accepted error can be found
9. repeat steps 1-4 until no more data subsets of can be w_i found
10. if (Ti does not contain any data examples of class w_i)
11. Generate the classifier.

Algorithm 2.1 PRISM Pseudocode

Database coverage pruning and its successors were one of the major breakthroughs in AC that can be employed successfully in induction learning as a late or post pruning method (Liu et al, 1998; Thabtah, et al, 2011). These pruning methods necessitate a positive data coverage per rule with a certain accuracy so the rule can be part of the classifier. A newer version of database coverage pruning was developed by Ayyat, et al, (2014) as a rule prediction measure. The authors proposed a method that considers the rule rank as a measure of goodness in addition to the number of similar rules sharing items in their body. These two parameters play a critical role in differentiating among available rules, especially in predicting test data class labels.

Recently, Elgibreen and Aksoy (2013) investigated a common problem in both the PRISM and RULES-IS algorithms, which is the trade-off between training time and classification accuracy during constructing the classifiers (Elgibreen & Aksoy, 2013). Moreover, the authors also highlighted performance deterioration of Covering methods when applied to incomplete datasets. Their result is a new induction algorithm that utilises the "Transfer Knowledge" approach to fill in missing attribute values, specifically the target attribute, in the training dataset before mining kicks in. "Transfer Knowledge" is basically building up a knowledge base based on learning curves via agents from different environments and previous learning experiences, which is generally used to fill in incomplete data examples (Ramon, et al, 2007). Experimental results against eight datasets reported in (Ramon, et al, 2007) revealed this the improved PRISM Covering algorithm consistently produced competitive classifiers when compared with other algorithms such as RULES-IS and PRISM.

One of the major obstacles DM algorithms face is the massive amount of data that are stored and scattered in different geographical locations. Decision makers are striving for an "on the fly" mining approach that processes very big databases simultaneously, hoping to facilitate planning decisions. Most of reported studies on parallelisation in classification were focused on decision trees (Stah and Bramer, 2014). However, Covering may lead to simple classifiers that are useful to decision makers, which suggest why Stahl and Bramer (2012) have investigated the big data problem by amending the PRISM algorithm to handle parallelisation., leading them to develop parallel learning method called Parallel Modular Classification Rule Induction (PMCRI). This strategy is a continuation of an early work by the same authors reported in 2008 which resulted in parallel PRISM (P-PRISM) algorithm (Stahl & Bramer, 2008).

P-PRISM algorithm aims to overcome PRISM's excessive computational process of testing the entire population of data attributes inside the training dataset. The parallelism involves sorting attribute values based on their frequency in the input dataset and class attribute. This means the algorithm will need only this information for processing the rules and therefore hold these results rather than the entire input data, which reduces usage of computing resources such as processing time and memory. The attribute values and their frequency are then distributed to clusters, and rules are generated from each. All rules are finally merged to form the classifier. Experiments using a replication of the Diabetes and Yeast datasets from the UCI repository (Litchman, 2013) reported in (Stahl & Bramer, 2008) with results showing that P-PRISM scales well in run time. However, a better approach to evaluate the parallelisation capability in P-PRISM is to utilise unstructured real data such as Bioinformatics or text mining where dimensionality is huge and the number of instances vary.

Since PRISM often suffers from over-fitting Stahl and Bramer (2014) developed an Ensemble Learning PRISM to minimise this risk. Ensemble Learning is an approach used in classification to improve the classifier's predictive accuracy by deriving multiple classifiers from any learning approach such as NN, Naïve Bayes, decision trees, etc, and then merging them to form a global classifier. The results derived from fifteen UCI datasets revealed that the Ensemble Learning model based on PRISM was able to generate results slightly better than classic PRISM algorithm (Stahl and Bramer, 2014). Moreover, a parallel version of the new model has also been implemented by the authors and tested with respect to training time. Results on run time showed that the parallel version scales well during the process of constructing the classifier.

PRISM successors have focused mainly on improving algorithm scalability. We have seen approaches such as P-PRISM that showed a promising research direction toward parallel DM using induction learning. Other approaches such as Ensemble Learning based PRISM were able to minimise over-fitting the training data by integrating Ensemble classifiers (weak learners) with J-pruning to reduce the number of rules. There is a need to further cut down irrelevant candidate items during rules construction in PRISM. This will have obviously a positive impact on the algorithm's efficiency in mining the rules and building the classifier. We believe that post-pruning is also essential in PRISM as it should increase its chance of being used as an intelligent solution and decision-making tool in practical applications. However, its use is limited because the classifiers produced by this algorithm family is large in regards to the numbers of rules.

Since PRISM relies primarily on an item's frequency to build rules; one promising solution to shrink the size of PRISM's classifier could be accomplished by having live item frequency during the mining phase. In particular, PRISM algorithm often employs the rule's 100% accuracy as a measure to generate the rule. This normally results in low data coverage by the rules. However, each candidate item that can be part of a rule body could be associated with its true frequency in the training dataset, which usually changes when a rule is generated. Recall that when a rule is derived by PRISM, all positive data linked with it is discarded and this may affect other items appearing in those discarded rows. While building the classifier, the true data representation for each item would decrease the search space and all insufficient candidate items would be deleted rather than stored as in PRISM. The overall benefit will be having less redundant rules in the classifiers. These classifiers can be seen as real decision support system tools since they would hold rules that are easy to maintain and use by decision makers. This pruning strategy will be implemented in our new eDRI algorithm discussed in Chapter 3 as a solution to decrease the classifier size in terms of the numbers of rules.

2.4.2 Incremental Reduced Error Pruning and Its Successors

Rule pruning in the separate-and-conquer approach enhances the predictive models' performance by minimising the number of errors on test data, in particular when the input data has noise (Franczak, 2000). The ultimate aim of rule pruning is to discard rules that over represent training data and under represent unseen data (test dataset). These types of rules, which misclassify instances, have limited training data coverage, or high similarity with portions of its rule do not perform well on test data. Consequently, early discarding of such rules during the classifier-building phase is preeminent and could enhance the general predictive performance of the models. Reduced Error Pruning (REP) (Quinlan, 1987), Incremental Reduced Error Pruning (IREP) (Fürnkranz and Widmer 1994), and Repeated Incremental Pruning to Produce Error Reduction (RIPPER) (Chen, 1995) are successful methods that have been used in rule learning models to accomplish the task of model generalisation by rule pruning (Quinlan, 1987; Pagallo & Haussler, 1990; Fürnkranz & Widmer, 1994; Cohen, 1995).

IREP and RIPPER are improved rule induction algorithms that minimise over-fitting resulting from overlearning the models. The aim is to extend the successful REP method so it can be used as a learning algorithm to produce efficient rule-based models. Most of the credit in rule induction algorithms based on REPs goes to Cohen at AT&T Bell Laboratories. Cohen's main goal was to design a fast rule induction algorithm that can scale well with C4.5 decision tree algorithms by using IREP pruning. Before discussing the RIPPER algorithm, both REP and IREP algorithms are presented.

2.4.2.1 *Reduced Error Pruning (REP)*

Initially, REP was designed to reduce the over-fitting problem of decision tree models by Ross Quinlan in 1987 so that predictive performance of the classifiers is enhanced. Later, REP was applied by Pagallo and Haussler (1990) and Brunk and Pazzani (1991) to produce decision lists. REP is a pruning method used in both rule induction and decision trees to reduce the error of a produced rule set. Therefore, REP can consider as a separate-and-conquer method that keeps pruning rules in the classifier until the classifier has an acceptable predictive accuracy. It should be noted that when REP is employed in decision tree algorithms like C4.5 or See5 the tree model is converted into the corresponding rules set before invoking it (Quinlan, 1993; Quinlan, 2002).

REP splits the training dataset into pruning and growing sets, with the learning algorithm deriving its rules from the growing set. The initial set of rules often over-fits the growing set, so REP applies a post rule pruning heuristic repeatedly to simplify the rules set by using rule pruning operators. This pruning heuristic evaluates each discovered rule of the pruning dataset, and the rule pruning operator that generates the smallest error is selected. To elaborate, each rule is evaluated against the pruning dataset, and then adjusts the rule body using operators in order to generate a new rule that ideally fits the pruning dataset. A rule pruning operator removes a condition (item) in the rule's body, or the whole rule in some cases. The rule pruning process terminates when removing an item increases the rule's error on the pruning dataset.

Experimental studies, such as Cohen (1995), Fürnkranz and Widmer (1994), Cohen (1993) and Brunk and Pazzani (1991) have shown enhancement in a models' predictive power when REP was adopted for rule pruning. Nevertheless, REP was criticised for being intensive resource demanding method since each rule must go through a thorough testing by pruning operators. This can be computationally costly when the input dataset is highly dimensional, which limits REP applicability.

2.4.2.2 *Incremental Reduced Error Pruning (IREP)*

In order to improve efficiency of the REP method, Fürnkranz and Widmer (1994) proposed Incremental REP (IREP) (Algorithm 2.2) as a rule-based learning method with a new pruning heuristic. Unlike REP, IREP introduced the idea of eliminating the generation phase of the initial rule set, thereby saving substantial data processing time and computing resources. Similar to REP, the training dataset gets split into pruning (33.33%) and growing (66.67%) sets. Then, IREP constructs greedily one rule at a time.

When a rule is put into a pruning test and passes that test, then its corresponding data instances in both the growing and pruning sets are discarded and the rule is derived. During rule pruning, IREP considers deleting conditions (items) from the rule's body and stops the pruning when no item deletion can enhance the following value given by equation (2.3).

$$v(Rule, prunePos, pruneNeg) \equiv \frac{p + (N - n)}{p + N} \qquad (2.3)$$

Input: A training dataset
Output: A ruleset classifier
1. procedure IREP(Pos,Neg)
2. begin
3. *Rule_set* ←{}
4. while Pos < > {} do

```
split (Pos,Neg) into (GrowPos,GrowNeg) and (PrunePos,Prune-
Neg)
R ←GrowRule(GrowPos,GrowNeg)
R ←PruneRule(Rule,PrunePos,PruneNeg)
if Error (R) on (PrunePos,PruneNeg) >= 50% then
    return Rule_set
else
    Rule_set ←R
    remove examples covered by Rule from (pos,Neg)
endif
endwhile
return Rule_set
end
```

Algorithm 2.2 Incremental Reduced Error Pruning procedure

Where p and n are the # of positive and negative instances covered by the pruned rule, and N is the total number of instances in the pruning set.

IREP intuitively eliminates the last condition added to the rule until the rule's accuracy cannot be improved, when the rule will be generated. Whenever a rule is derived, all remaining uncovered instances in the training dataset are re-partitioned. The same process is repeated until no more positive instances are left, or the last rule derived had an unacceptable level of error. In other words, the termination occurs when the error of the last pruned rule is larger than that of the empty rule. When this occurs, the discovered pruned rules are returned as the classifier. Experimental studies by Fürnkranz and Widmer (1994) and Cohen (1995) shown that IREP is highly competitive with regards to classification accuracy compared to REP, and more importantly, faster than REP in producing the rules set.

```
Input: A grow dataset
Output: A new rule: R
procedure Grow Rule (GrowSet)
begin
R ←{}
do
Globa Best Split ←{}
for each variable in GrowSet do
NewSplit ←FINDSPLIT (var iable, GrowSet)
```

```
if NewSplit is better than Global_Best_Split
    Global Best Split ←NewSplit
R ←R + Global_Best_Split
GrowSet ←NOTCOVERED(R, GrowRule)
until no errors on GrowSet
remove examples covered by Rule from (pos,Neg)
end
```

Algorithm 2.3 Growing Rule Function used in REP, IREP and RIPPER algorithms

2.4.2.3 *RIPPER and Other Alternative Algorithms*

To improve performance of the REP and IREP methods, the RIPPER algorithm was proposed by Cohen (1995). RIPPER was shown to have a higher predictive accuracy and less complexity when contrasted with IREP. RIPPER initially employed a divide-and-conquer strategy integrated with a Covering search heuristic during the process of rule induction. Specifically, after a rule is found, similar to IREP, RIPPER prunes the rule by considering removal of some of its conditions and evaluating the resultant error rate. RIPPER was designed to handle multi-class datasets and binary classification data, unlike IREP which was developed primarily for datasets with two possible class values. This is a bonus for RIPPER in mining different real world application data. Algorithm 2.3 depicts the function of growing the rule used by the REP, IREP, and RIPPER algorithms.

One of the important enhancements over REP is the optimisation phase added. In this phase, RIPPER amends or removes the rules set in order to improve the predictive accuracy of the constructed model. Particularly, and for each rule, RIPPER creates two alternative rules: the replacement rule and the revision rule. The latter is formed by adding literal conditions to the rule body until the error on the rule increases. However, the replacement rule is formed by growing a rule from an empty set and removing literals from the rule body until the error on the entire rule set gets minimised. The decision whether to include the original rule, its replacement rule, or its revision rule is made based on evaluating them against the minim description length principle (MDL) (Witten & Frank, 2005).

Dian, et al. (2004) proposed IREP++ to further enhance runtime of the IREP algorithm while reducing the number of items a rule can hold, thereby generating less complex models. The authors used the Gini Index metric as the basis for the rule growing function procedure. The Gini Index evaluates data impurity based on the following Equation 2.4.

$$I(S) \equiv \frac{ps}{(ps+ns)} \cdot \frac{ns}{(ns+ps)} \tag{2.4}$$

where *ps* is the number of positive instance and *ns* is the number of negative instances.

IREP++ splits the training dataset, similarly to IREP, into growing and pruning sets. Then during the rule building process, IREP++ uses the data split that leads to the largest reduction in Gini Index based on Equation 2.5 below,

$$\Delta I = I(S) - \frac{ps_1 + ns_1}{(ps+ns)} I(S_1) - \frac{ps_2 + ns_2}{(ps+ns)} I(S_2) \tag{2.5}$$

When a rule is about to be pruned, IREP++ adopts the RIPPER rule pruning strategy in which any sequence of literals (conditions) added to the rule is considered (Cohen, 1995). The stopping metric for rule pruning used.

$$M(p,n) \equiv \frac{p-n}{p+n} \tag{2.6}$$

where p and n are the number of positive and negative instances covered by the pruned rule. P and N are the number of positive and negative instances in the entire pruning set.

Cohen and Singer (1999) have developed a variation of the RIPPER algorithm called SLIPPER based on Boosting learning. SLIPPER improved the predictive models slightly when compared with RIPPER because of the Boosting strategy implemented during the process of rule construction. In Boosting a rule, each successive rule classifies training data instances that were incorrectly classified by the rule's predecessors (rules already generated). This new rule learning strategy has resulted in a different classifier to RIPPER at least on the sample of datasets used in Cohen and Singer (1999). However, the SLIPPER algorithm suffers from the high complexity caused by the Boosting mechanism, which makes it slower than traditional rule induction algorithms.

Experimental results against nine datasets showed that IREP++ scales well with the IREP and RIPPER algorithms with respect to run time.

2.4.3 Rough Set Theory Induction Algorithms

Pawlak introduced Rough set Theory as a mathematical solution to uncertainties and inconsistencies in data analysis (Pawlak, 1982). Many datasets contain incomplete data instances, missing values, miscoding or mislabelling. Therefore, one may not be able to classify instances accurately in light of incomplete datasets. Rough set theory introduces a solution by utilising approximations referred to as lower and upper approximation approaches (Pawlak, 1991). One may utilise rough set theory to generate rules in cases where the input data is incomplete.

Unlike other approaches used to deal with inconsistencies and uncertainties in input data, rough set theory does not aggregate nor correct the missing pieces. It relies on lower and upper approximations to compute concepts in inducing rules. Therefore, one may think of this approach as dealing with consistent data, since it does not alter the input structure used in rule induction. The output generated by rough set theory can be categorised into two classes of rules: certain and approximate, contingent on whether lower or upper approximations are used (Pawlak, et al, 1995).

Few algorithms have been developed to induce rules based on the rough sets' approach. Learning from Examples Based on Rough Sets (LERS) is a good example of the systems used to induce rules based on rough sets theory (Pawlak, 1991). LERS has many variants to induce rules, all using the aforementioned framework that utilises approximation to generate accurate rules. LERs is a system used to induce a set of rules based on examples presented in a decision table format, and may be used to classify new examples based on the learned set of rules (Pawlak, 1982). The decision table contains input data collected from a certain field such as banking, nursing, manufacturing or sociology. Variables are included in the table (symptoms or test outcomes), along with the decision and dependent variables. The decision table is similar to the training dataset.

LERS simply searches for regularities in the table. The output of LERS is a set of rules, used to classify or predict new instances. Initially, LERS uses data instances for inducing classification rules from the decision table using a Covering procedure described later in this sub-section (Algorithm 2.4). LERS, similar to other algorithms, attempts to find a minimal description of concepts presented in the data by retaining positive examples while excluding negative examples. Therefore, the rules generated by LERS are referred to as minimal discriminant concepts.

LERS is capable of dealing with uncertainty in data. Missing values and inconsistency are the two main problems constituting this uncertainty in a given dataset. LERS has been implemented to process continuous variables from inconsistent data to produce rules; LERS is applied by many decision support systems such as LEM2 (Pawlak, et al, 1995). The primary benefit of rough set theory is that it does not require additional information about the presented data. Rough set theory does not remove the inconsistencies, however, but rather finds approximations to the concept (Pawlak, 1991). The remaining part of this section describes the Covering procedure of LERS and its variations while introducing the mathematical notations.

Let X be subset of the set Y of all variables and X is nonempty. Let W be the set of all instances, the indiscernible relation $IND(X)$ is a relation on W defined for $a, b \in W$ by $(a, b) \in IND(X)$ if and only if for both a and b the values for all variables from X are similar.

```
Input: Set of all variables, {c}* on W
Output: A global Covering R
do begin
compute X split
X ←P
R ←θ
if X ≤{c}*
  begin
for each variable x in X do
 begin
   D ←P −{x}
   calculate split D*
   if D*
     P ←D
     end // for
   R ←P
  end//if
end
```

Algorithm 2.4 Single global Covering function of LERs and its variations (Pawlak, 1991)

The indiscernible relation is an equivalence relation, and the classes of IND (X) are named elementary sets of X. For instance, from Table 2.2 (from Grzymala-Busse, 2010), and X= {Temperature, Headache}, elementary sets of IND (X) are {{1}, {2}, {3,7}, {4}, {5,6}}. All X-elementary sets have a family X^*. For a class c, c depends on X if $X^* \leq \{c\}^*$. A Covering (reduct) of {c} is a subset of *X* of *Y* such that {c} depends on *X* and *X* is minimal in *Y*. One can obtain the Covering of {c} by comparing splits of X* with {c}*.

Table. 2.2 Sample dataset (Grzymala-Busse, 2010)

Instance ID	Temperature	Headache	Weakness	Nausea	Class
1	Very high	Y	Y	N	Yes
2	High	Y	N	Y	Yes
3	Normal	N	N	N	No
4	Normal	Y	Y	Y	Yes
5	High	N	Y	N	Yes
6	High	N	N	N	No
7	Normal	N	Y	N	No

The Covering methodology (see Algorithm 2.4) used in LERS components such as LEM1 or LEM2 are calculated based on the eliminating condition procedure developed by Michalski (1983). Using this procedure, the condition within a rule such as R1 is scanned from R1's body starting with the left-hand most condition moving to the right. Whenever a condition is removed, we check the decision table to evaluate whether the new simplified rule is violating the consistency of the discriminant description (Grzymala-Busse, 2010).

2.4.4 CN2 Algorithm

CN2 is a classification algorithm designed to induce a set of rules from datasets associated with duplication (Witten & Frank, 2005). The original CN2 algorithm (See Algorithm 2.5) was initially designed in 1989 and then improved in 1991 (Clark & Niblett, 1989; Clark & Boswell, 1991). The algorithm traditionally used Entropy for its search heuristic and was able to produce a list of ordered rules. Entropy indicates variable purity in splitting data instances with respect to the class label. The purer a variable can split, the training instances with reference to the target attribute values (class variable), the higher the variable score. Many have shown that the algorithms' performance increases with few modifications (Quinlan, 1993; Abdelhamid and Thabtah, 2014). For instance, using a Laplacian error estimate as the heuristic may improve the results of the models (Witten & Frank, 2005).

CN2 is composed of two main phases: a search and a control phase. The search is composed of a beam search with the objective of reaching to a good rule, while the control phase is composed of the reiteration and execution of the search algorithm part. During the search phase, the algorithm looks for the best rules it can generate. One possible evaluative criterion used is the accuracy rate of rules given the training data. CN2 follows the learning approach that first identifies the rule's body then the class, rather than class then conditions as in the REP, IREP and RIPPER algorithms.

In 1995, an improved search algorithm called Optimized Pruning for Unordered Search (OPUS) was proposed by Webb (1995). OPUS presents a solution for the heuristic search problem in ML predictive tasks. OPUS is a fast heuristic for searching the data space for conjunctive rules. Usually, heuristic search methods do not guarantee generation of the target relation. However, OPUS presents an alternative search strategy, based on admissible search, that promises production of the target relations if they exist. OPUS method permits efficient admissible searches for search spaces in which the order of application of search operators is not important. This is permitted by the use of domain specific pruning of rule to provide a narrower search space. OPUS uses a branch and bound search strategy that guarantees that no two equal nodes in the search space are visited.

```
Input: Set of all variables (training dataset-T)
Output: Rules set (R_S)
function CN2 (T)
begin
R_S ← θ // initialising the rule set to be empty
do
   begin
   find_Best_Variable(T) // Entropy metric is used to measure the best variable
   temp ← variable_subset_T // Inserting the best variable data examples of the variable in a temp
   R ← Var_value // Adding a variable value to the rule
   T ← T − temp //removing the rule covered data from the training dataset
   R ← CommonClass(Temp) // finding the most common class in the rule's data examples
   until no adding a variable is not satisfies the minimum condition
end
R_S ← R // adding the rule into the rule set
end
```

Algorithm 2.5 CN2 original rule generation function (Clark and Niblett, 1989)

It also optimises the influence of pruning by organising the search process.

2.4.5 First Order Inductive Learner (FOIL)

First Order Inductive Learner (FOIL) builds learning systems that utilise propositional value attribute language to describe entities and classification rules, as well as systems that generate first order logic classification rules (Quinlan, 1990). FOIL originated from developments based on other classification methods including the divide-and-conquer approach ID3 (Quinlan, 1979). The FOIL algorithm aims to generate statements that describe a specific relation by itself and other relations in the input data. FOIL shifts from the explicit approach of representing the relation with a set of tuples of a particular collection of constants to a general functional form applied to different collections of constants.

FOIL rule generation functions (Algorithm 2.6) devise statements one at a time to describe a relation, and then looks for the subsequent one. It then uses an information based heuristic to generate a general functional set of statements. Thus, FOIL begins with a set of tuples of constants instead of the variable-values structure. Each tuple corresponds to a value assigned to a variable. They are then indicated to be either positive or negative, depending on whether or not they correspond to the target relation. More details on the mathematical formulas of FOIL can be found in Quinlan (1990).

```
Input: Set of all variables (training dataset-T)
Output: Rules set (R_S): first order predicate logic
Foil (T)
begin
   R_S ← θ
   Set P_S be the positive set of data examples (data belong to
   class C)
   Set C class be the predicate to be learned
for each p_s instance in P_S do
begin
   let N be the negative data examples
   R ←If nothing, then C // creating an empty rule
   find_Best_Variable(T) // FOIL gain metric is used to measure
   the best variable
   R ←Var_value // Adding a variable value to the rule's body
```

12. end
13. delete from P_S data examples which match the rule's body
14. until N is empty do
15. select a literal L
16. append L to Rule's Body
17. delete from N data examples that do not match L
18. end

Algorithm 2.6 FOIL algorithm (Quinlan, 1990)

2.4.6 Instance Based Learning Rule Classifiers

Domingos (1996a) proposed a new rule induction approach called RISE, linking it with instance-based learning in classification (Cover, et al, 1967; Aha, et al, 1991). The RISE approach does not contain a global procedure, but it considers all algorithms (growing, pruning, etc.) as sub procedures. RISE generates a classifier that can be considered a rule induction set, as well as a nearest neighbour method. The author refers to RISE as a unified rather than combined approach that looks for the rule in a specific-to-general manner. It begins with one rule per classified training instance. In doing so, it circumvents the few limitations presented by the Covering approach to classification. RISE evaluates the accuracy of a set of rules all at once in a global fashion. It then uses a best match framework empirical evaluation when assigning class labels to test data. Experimental studies revealed that RISE outperforms its parent approaches, such as CN2. In particular, Domingos in (1996a) Domingos (1996b) showed that RISE achieved slightly better classification accuracy than classic rule learning algorithms. For large datasets, though, RISE suffers from the long time spent discovering and pruning the rules, which may obviously limit its applicability in big data applications.

Gora and Wojna (2002) proposed a combined approach similar to RISE, called RIONA, for rule induction that incorporates elements of rule-based classification techniques. In RIONA, decisions are not based on the set of all rules or matching test cases, but on a restriction of rules applicable neighbourhood test cases. The size and optimisation of the restriction is automatically produced. Empirical testing of RIONA indicates that it is sufficient to achieve acceptable levels of classification accuracy by specifying a narrow neighbourhood.

RIONA and RISE can be considered learning approaches based on integrating rule-based classification and the nearest neighbour method. Yet RIONA is different to RISE because it considers all minimal rules (rules with a limited number of attribute values with larger data coverage). This suggests that the class assignment decision should use a class that is more frequent in the minimal rules matching a test case. The primary idea of RIONA is that the support set is constrained to the neighbourhood of a test case. The neighbourhood is composed of training data within a specified distance from a test data, or a quantity of data closest to a test data similar to the K Nearest Neighbour algorithm (Cost & Salzberg, 1993). The ideal size of a neighbourhood applied to classifying a test case is learned during the training phase. One distinguishing feature of the RIONA algorithm is cutting the search time for the minimal rules set because observations are taken from the neighbourhood rather than from the complete training phase dataset. Moreover, rules are extracted by RIONA during the classification phase, and each produced rule is associated with particular test data. A rule descriptor is expressed by the test data attribute values.

2.4.7 One Rule and RIDOR Algorithms

One of the simplest, yet acceptable, algorithms that produce one rule classifiers is SimpleRule /OneRule (Holte, 1993). OneRule initially scans the input dataset and looks for the pair of <attribute value, class> with the least error rate on the training dataset. This pair is produced as a rule classifier. Algorithm 2.7 depicts the description of OneRule algorithm steps from Witten & Frank (2005). Normally, the OneRule algorithm creates a frequency table for each attribute value in the dataset in order to calculate the errors for that attribute value when associated with a class label.

A more advanced method that generates rules as exceptions is called the Ripple Down Rule learner (RIDOR) proposed by Gaines and Compton (1995). RIDOR normally builds a default rule, then repeatedly employs the IREP method to generate exceptions with the least number of errors and then identify exceptions from each discovered exception by reviewing subsets of the training dataset.

Input: Training dataset T

Output: One rule classifier

1. for each variable value, create a rule by:
2. find the largest frequency class label
3. assign that class to the variable value
4. compute the number errors of the rule
5. choose the variable value with the least errors.

Algorithm 2.7 OneRule Pseudocode

```
Input : A training dataset T with target class Rt that contains P positive
and N negative data examples
Output : A set of rules
Ridor (T, Rt)
R ←θ
If |Rt|< Minimum_support
R ←θ
Mark R active
Do
    Find a rule:r in active relation
    Learn except and if not branch
    Set relation of r to active
    R ←R + r
    X ←X + r
Until X= θ
Set all active relations into inactive
Return R
end
```

Algorithm 2.8 Ridor algorithm pseudocode (Gaines and Compton, 1995)

Good exceptions are those derived by each exception in tree like expanding. The exceptions denote classification rules that forecast class labels other than the default class rule. At the heart of this algorithm is IREP since it is implemented to produce rules from exceptions. Algorithm 2.8 shows the RIDOR algorithm pseudocode.

2.4.8 Hybrid Classification Rules

A classification method that integrates divide-and-conquer with separate-and-conquer, called PART, was proposed in Frank & Witten (1998). PART uses information gain and decision tree pruning to build partial decision trees that are converted into rules sets. PART constructs partial decision trees from the input dataset and then a rule induction strategy is invoked to produce the rules set. Thus, PART makes a rule from the sub-tree and transforms each path from the root node to the leaf node with the largest data frequency into a candidate rule, then discards the sub-tree and its corresponding positive data from the training dataset. PART repeats this process until the dataset becomes empty, or no rule with acceptable error can be produced.

An integration of fuzzy logic and rule induction was implemented in an algorithm called Fuzzy Unordered Rule Induction Algorithm (FURIA) (Hahn & Hllermeier, 2009). FURIA enhanced the RIPPER algorithm maintains its main advantages by including its optimisation procedure, pruning and prediction methods, and comprehensive rules set. FURIA adds into RIPPER fuzzy rules rather the classic rules induced, so unordered rule sets are presented instead of rule lists. In addition, FURIA illustrated a new rule stretching method to deal with data examples that have no rule coverage, thus minimising the use of default class rule and reducing the number of misclassifications. Results obtained from experimentations revealed that FURIA outperformed RIPPER and C4.5 decision tree algorithms with respect to predictive power (Hahn & Hllermeier, 2009).

The way FURIA discovers the unordered rules set per class label is based on the one versus the rest strategy. Unlike conventional rule induction approaches that produce decision lists, models created by FURIA may not be comprehensive which can cause some training data not to be covered by the fuzzy rules which is problematic. To treat this case, FURIA is empowered with an efficient rule stretching method adopted from Eineborg and Boström (2001). This stretching method generalises the rules set until the data example is classified.

2.5 Other Common Classical Non-Rule Approaches

In this sub-section, we briefly review common ML methods such as Probabilistic, Boosting, SVM, NN, and AC. This has been done since some of these algorithms have been employed in chapters 4 and 5 for comparison purposes.

2.5.1 Probabilistic Models

One of the simplest methods to come up with a decision based on a labelled set of variables is to utilise probabilistic methods. In classification problems, the Naïve Bayes classification method is based on the simple and naïve assumption that attributes in the training dataset are independent and have no correlation (Duda & Hart, 1973). This method was initially developed in the 1950s and later gained popularity because of its efficient performance and applicability to classification problems including text mining and image classification among others (Hand & Yu, 2001). In Naïve Bayes, any variable value is assumed to be totally independent from other variables' values in the training dataset.

When test data is about to be classified, the Naïve Bayes algorithm computes the likelihood of the attribute values in the test case from the training dataset with respect to each class label. In other words, each attribute values' probability in regards to each class is calculated. The test data attribute values probabilities are then multiplied with each other and the class distribution probability computed. Test data is then assigned the higher probability outcome based on these aforementioned calculations (see Equation 2.7). Naïve Bayes is a highly scalable algorithm, and can produce the classification decision quickly when compared against other classification approaches (Ng & Jordan, 2002). Moreover, the predictive power of the classifiers generated by probabilistic algorithms is acceptable when compared with Boosting and divide-and-conquer algorithms (Caruana & Niculescu-Mizil, 2006). One clear disadvantage of Naïve Bayes, however, is its independent assumption since attributes are usually correlated in reality.

Given a test case represented as a vector X = (x_1, x_2, …, x_n) where each x is an attribute value to be classified, using Bayes Theorem, the conditional probability can be computed as:

$$P(C_n \mid X) = \frac{P(C_n).P(X \mid C_n)}{P(X)} \qquad (2.7)$$

The test data X will be assigned the class C_n with the largest conditional probability value $P(C_n \mid X)$.

2.5.2 Association Rule-based Classification

Associative Classification (AC) is an approach that employs association rule methods to generate rule-based classification systems from supervised learning datasets (Liu, et al., 1998; Thabtah, et al., 2005; Abdelhamid, et al., 2016). This approach normally relies on user minimum support and minimum confidence. Minimum support of a feature value is how many times that feature value has appeared in the training dataset whereas the frequency of the feature value plus the target class together denote the confidence of that feature value.

The aim of the AC algorithm is to discover all frequent feature values, and from these the classification system is built in which any frequent feature value with confidence larger than the minimum confidence becomes knowledge. One advantage of AC is the low error rate systems derived, but a noticeable drawback is the very large number of rules derived which limits its usage in real applications. Another noticeable drawback of AC is the time needed to learn and evaluate the knowledge. This is because a typical AC method, like CBA, requires multiple data scans to discover the knowledge (Abdelhamid & Thabtah, 2014). A number of comparative studies have shown that AC is more accurate than rule induction, decision trees, and probabilistic models (Abdelhamid, 2015; Thabtah, et al, 2004). Nevertheless, associative classifiers demanded high computing resources and may crash when the support is set to a low value. Hence, they are not suitable for large datasets.

2.5.3 Support Vector Machine

Support vector machine (SVM) is a common approach that was initially developed to improve the performance of current classification methods, particularly classification accuracy (Boser, et al., 1992). SVM relies on decision planes, which split a set of data objects using computed class memberships. The logic behind SVM learning methodology is that it rearranges data instances within the problem space utilising quantitative functions known as kernels. A kernel is a function that measures similarity between data objects with respect to class labels (Joachims, 1999). Normally kernels are given by the domain experts to the SVM algorithm and used during the classification process. The entire procedure of changing the orientation or data instances is referred to as transformation. Experimentation showed that SVM methods normally generate better classifiers with respect to accuracy than traditional classification approaches such as decision trees and rule induction (Joachims, 1999; Shawe-Taylor & Sun, 2011).

2.5.4 Boosting

Boosting is a supervised learning approach in ML that was mainly designed to minimise the problem of bias in classification data (Schapire & Singer, 2000). The origin of the Boosting learning approach was a question of whether a group of weak learners can make one strong learner. A weak learner is basically a model/classifier that can help classify unseen data better than random classification (Freund & Schapire, 1997). Therefore, Boosting is based on the idea of repeatedly improving the accuracy of a classifier by merging weak and inaccurate rules. Typically, after boosting has derived the weak learners it reweighs the training data accordingly, where the performance is then boosted (Schapire & Singer, 2000`).

The majority of current Boosting algorithms like AdaBoost repeatedly learn weak classifiers then add them to make a final single classifier (Freund & Schapire, 1997). When the classifiers are added, they are weighted based on the accuracies derived from them and, when a particular weak classifier is added, the data gets reweighted. In the data re-weighing, instances that were incorrectly classified gain higher weights and those that were correctly classified drop their weights. This decrement and increment in training data weights enable near future weak classifiers to focus more on data instances that have been misclassified by previously weak classifiers.

2.5.5 Neural Network

An artificial neural network (ANN) often consists of an interconnected processing components (neurons) that converts a set of inputs into desired output (Witten & Frank, 2005). The strength of ANN originates from the hidden neurons' nonlinearity. The output, is normally dependant on certain features of the components and weights linked with their interconnections. Usually the ANN classification model is shaped after the training phase on the input dataset in which information about the weights and features related to the network's components heavily impacts the performance quality of the formed predictive system. The most used function of ANN is called the sigmoid function (Mohammad, et al, 2012). During formation of the NN predictive model, weights keep amending to derive accuracy closer to the desire value specified by the end-user.

2.6 Chapter Summary

Classification in ML is a vital task that commonly is accomplished by applying various learning approaches. This chapter has reviewed common rule-based approaches in classification, especially those that are based on separate-and-conquer including rule induction and Covering approaches. In particular, we have critically analysed algorithms that produce rules such as PRISM and its successors, REP, IREP, RIDOR, RIPPER, RIONA, FURIA, and PART among others. Furthermore, we shed light on divide-and-conquer and other non-rule-based classification systems such as decision trees, probabilistic, Boosting, SVM, etc.

In the next chapter we propose our novel rule learning approach, which we called enhanced Dynamic Rule Induction (eDRI) and highlight its major steps, strengths, and weaknesses as well.

Chapter 3

A New Dynamic Rule Induction Approach

3.1 Introduction

As aforementioned, Covering algorithms such as PRISM usually produce "If-Then" classifiers with comparable predictive performance to other traditional classification approaches such as decision trees and associative classification. Thus, these classifiers are favourable for users because they can be useful tools in decision making. Nevertheless, Covering methods, including PRISM and its successors, suffer from a number of drawbacks such as the large number of rules derived. This can be burdensome, especially when input data is largely dimensional. Therefore, pruning unnecessary rules becomes essential for the success of this type of algorithm.

This chapter proposes a new rule-based classification algorithm that reduces the search space for candidate rules by pruning early on any irrelevant items during the process of building the classifier. Whenever a rule is generated, this algorithm updates the candidate items' frequency to reflect the discarded data examples associated with the rules derived and thus avoids unnecessary repetitive data scans. This methodology makes item frequencies dynamic rather than static and it ensures that irrelevant rules are deleted in preliminary stages when they do not hold enough data representation. The major benefit will be the production of a concise set of rules, which facilitates their management by users and therefore appealing as decision making tools.

This chapter proposes a classification method called enhanced Dynamic Rule Induction (eDRI). After the initial scan of the training dataset, eDRI discovers the rule one-by-one. It primarily uses a threshold value to limit the search space for rules by discarding any items with insufficient data representation as early as possible. These items are called weak items since they do not hold sufficient statistical significance. eDRI appends the largest item with respect to accuracy into the current rule and continues appending items until the current rule cannot improve any further, i.e. its strength is above the Rule_Strength threshold, which is used to separate acceptable from unacceptable rules.

Once a rule is produced, eDRI removes all of its training data instances and amends all candidate items' frequency that appeared within the deleted instances of the generated rule. This approach leads to a natural pruning during the rule discovery phase and results in a more realistic classifier with lower number of rules. This is unlike induction algorithms that employ excessive divide-and-conquer pruning per rule after the rule generation phase, which consequently makes the training phase slow. In addition, the proposed eDRI algorithm limits the use of the default class rule by generating rules with accuracy below 100% accuaracy. Often, other algorithms ignore these rules because they do not have zero error. These rules are though used only during the class prediction phase instead of the default class rule, when there is no rule with 100% accuracy to classify a test data.

The proposed eDRI algorithm has been implemented in Java within the Waikato Environment for Knowledge Analysis (WEKA) environment, making it available to be utilised by different types of users such as managers, researchers, students in their own research studies and experimentations (Hall, et al., 2009). This ML platform was initially developed at the University of Waikato in New Zealand and contributors from all over the world who are interested in data analysis have since contributed specific methods to it.

Experimental results using real data from the security domain (Chapter 5) as well as different classification datasets from the University of California Irvine (UCI) (Chapter 4) reveal that the proposed eDRI algorithm is competitive in regards to classification accuracy when compared to known rule-based algorithms such as PRISM, OneRule, ConjunctiveRule, Decision Trees (C4.5), RIPPER, and hybrid approaches such as PART and Boosting. Moreover, classifiers produced by our eDRI algorithm are smaller in size, which increases their usability in practical applications. Further details on the implementation of eDRI and the experimentations carried out are provided in Chapter 4.

The rest of this chapter is structured as follows. Section 3.2 describe in detail the different phases of eDRI along with corresponding pseudocodes. Section 2.3 presents a thorough example using a dataset to show the pros and cons of eDRI and the classic Covering algorithm PRISM. Section 2.4 is then devoted to the distinguished features of eDRI, and lastly, the chapter summary is given in Section 2.5.

3.2 Enhanced Dynamic Rule Induction Algorithm (eDRI)

The proposed eDRI algorith (Algorithm 3.1) has two primary phases: rule production and class assignment of test data. In phase 1, eDRI produces rules for each available class label in the training dataset that has accuracy greater than or equal to *Rule_Strength* parameter. This parameter aims to generate near perfect rules alongside rules with 100% accuracy, like in classic PRISM, and it is used within eDRI to differentiate among acceptable and unacceptable rules. eDRI's learning procedure ensures the production of rules that have no error as well as rules that survive the Rule_Strength parameter, i.e. rules with expected accuracy greater than or equal to Rule_Strength parameter. The algorithm terminates building the classifier when no more rules with an acceptable accuracy are found, or when the training dataset becomes empty. When this occurs, all rules are merged together to form the classifier.

Another important parameter is utilised in phase 1 of the eDRI algorithm in phase 1, which aims to minimise the search space of items. The parameter *freq* is utilised primarily to differentiate between strong items, having a high number of occurrences in the training dataset, and weak items. This helps in removing weak items, i.e. items with low frequencies (frequency below the *freq* threshold), as early as possible and thereby saving computing resources as well as ensuring only strong items can be part of a rule's body. Weak items are maintained in PRISM in the hope of producing 100% accuracy rules. However, this is seen as a major deficiency because the classifier may end having too many specific rules covering small portions of the training dataset (Abdelhamid and Thabtah, 2014). On the other hand, because eDRI discards weak items it generates highly predictive general rules with large data coverage.

```
    Input: Training dataset T, Rule_Strength, Frequency (freq)
    thresholds
    Output: A classifier that consists of If-Then rules
Phase (1)
1.    data transformation and discretisation (if necessary)
        (See Figure 3.2 for rule inducing phase)
2.    for each Attribute in T Do
3.      make an empty rule, r_j: If Empty then Empty
4.      calculate Attx accuracy in T_i p(class_i = i | Attx);
5.      append the Attx with the largest accuracy (class_i = i | Attx) to
        the body of r_j
```

6. repeat steps 1.1-1.3 until r_j has either

7. 100% accuracy or no longer can be improved

8. r_j accuracy >= Rule_strength

9. generate r_j

10. discard all data examples from T_i that match r_j's body

11. update the frequency of all impacted candidate items to reflect step 3

12. repeat steps 1-4 until no items with acceptable frequency is found or no more data can be found

13. If (T does not contain any data examples)

14. generate the classifier

15. produce a default rule

Phase 2

16. predict test data (Algorithm 3.3)

Algorithm 3.1 eDRI Pseudocode

The eDRI algorithm generates all rules in phase 1. Whenever a rule is generated, training data examples linked with it are deleted and the frequencies of those waiting to be added containing strong items that appear in the deleted data are instantly updated. This update process involves decreasing their frequencies, thus some of these items may no longer become strong and are discarded, which can be seen as a quality assurance measure because it does not rely on the original frequency of items computed initially from the training dataset. Rather, we have a dynamic dataset that reduces its cardinality whenever a rule is generated. This reflects the true data coverage per rule in the final classifier and it contributes in the production of controllable classifiers that can be used as decision making tools in different application domains. Also, eDRI algorithm data processing procedure is more efficient as a result of this update process. Indeed, other advanced induction learners, such as RIPPER for example, scan the entire updated training dataset each time a rule is derived, whereas eDRI avoids this expensive data processing by actively updating item frequencies.

Whenever a rule is derived, its associated training data samples are removed; the frequency of each affected item is continuously updated during rule learning until the rule production phase is terminated. Having said this, eDRI is a special algorithm because it does not allow items inside rules to share training data examples, therefore generating classifiers that do not depend on a static training dataset. Rather, eDRI updates the training data and item frequencies each time a rule is formed.

In phase 2 of eDRI, rules derived are used to guess the type of class for test cases. Our algorithm assumes that the attributes inside the training dataset are nominal, and any continuous attribute must be discretised before the rule inducing phase starts. Finally, missing values are treated as any other values in the training dataset.

3.2.1 Data Representation and Data Processing

Before the eDRI starts searching for knowledge (rules), the training dataset is transformed into a data structure of the form <Item, class, Line#'s / row IDs>. The item and class are represented by <ColumnID, RowID>, where the first column and row numbers that the item/class occurs in is the training data set denoting the item/class. This data representation has been adopted from Thabtah and Hammoud (2013) because it minimises the counting of items and it also decreases the number of data scans the algorithm must go through during data processing. Another primary advantage of using this data formatting method is that there is no attribute value frequency counting after iteration 1, scanning the training dataset. This is because the algorithm stores both the attribute values and class locations of the training data set inside a special data structure, <Item, class>, which is utilised to locate the frequency of each item by just taking its size. This is a simple process that normally reduces the number of passes over the training dataset.

In eDRI, each data instance in the training dataset is linked to a unique integer value (Item number). The first row number where the instance occurs within the training dataset is called Rowld, while Columnlds is used for the variable (attribute) number in the training dataset which generates the item's values. Attributes are saved from left to right where the first one being denoted by "0," the second one by "1," and so forth. This data format helps in the data processing phase, especially while the algorithm is building the rules and performing computations for all candidate items in the training dataset in order to determine the survived items. Table 3.1b shows a data format example after transforming the training dataset of Table 3.1a. In Table 3.1b, (0)0 corresponds to item "X", and (0)1 correspond to item "Z," and so forth. This numbering method eases the mining process and reduces needed computing resources, especially training time during calculating and updating item frequencies, as it will be seen later on in Chapter 4.

Table 3.1A Training dataset sample

ROWID	Variables			Class
0	X	Y	Z	C1
1	Z	Y	Z	C1
2	Z	W	Z	C2
3	Z	W	Z	C3
4	X	Y	X	C2
5	X	W	X	C3
6	Z	W	X	C3
7	Z	Y	W	C3
8	X	Y	X	C3

Table 3.1 B Transformation of Table 3.1A

ROWID: Class	Variables		
0:0	(0)0	(1)0	(2)0
1:0	(0)1	(1)0	(2)0
2:2	(0)1	(1)2	(2)0
3:3	(0)1	(1)2	(2)0
4:2	(0)0	(1)0	(2)4
5:3	(0)0	(1)2	(2)4
6:3	(0)1	(1)2	(2)4
7:3	(0)1	(1)0	(2)7
8:3	(0)0	(1)0	(2)4

The eDRI algorithm assumes that all variables or attributes are categorical, meaning they have a finite set of possible values. Any continuous attribute must therefore be discretised using a particular discretisation method such as the entropy based one. For datasets with noise, eDRI takes advantage of the filtering procedures available in WEKA, with "ReplaceMissingValues" filter being a possible one. First, missing values in the training dataset should be replaced or treated as any other possible values. This filtering procedure was modified and embedded within eDRI in WEKA platform.

3.2.2 Rule Discovery and Classifier Building

During the rule discovery phase and earlier after transforming the training dataset into the designated format, the eDRI algorithm scans the training dataset to record items and their frequencies as described. All <item, class> appearances are counted and stored in a temporary data structure. Any item that appears with a class in the training dataset accompanied by inadequate frequency is discarded as encountered. This minimises the search space of all candidate items and ensures that only strong items can be part of the classifier's rules. Table 3.2 shows the candidate items obtained from Table 3.1a by eDRI, where the bold ones are strong items assuming that the freq threshold is set to 2.

All non-bold items of Table 3.2 will be removed by eDRI after passing over the dataset. From Table 3.2 it is obvious that there is one weak and six strong items of size one.

It is important to mention that unlike other rule-based classifications that are not covering-based, such as associative classification that require joining items and counting beginning with strong items of size 1 until n, eDRI only keeps items of size 1 during its rule discovery process. Therefore, eDRI relaxes one of the major burdens of associative classification as well as the association rule mining method, which uses the candidate generation function. This function normally necessitates passing over the training dataset many times to discover frequent items of size 2, 3, 4, to n.

Table 3.2 Items occurrences after initial scan by eDRI

Variable	Variable Frequency
(0) 0	**0:0, 4:2, 5:3, 8:3**
(0) 1	**1:0, 2:2, 3:3, 6:3, 7:3**
(1) 0	**0:0, 1:0, 4:2, 7:3, 8:3**
(1) 2	**2:2, 3:3, 5:3, 6:6**
(2) 0	**0:0, 1:0, 2:2, 3:3**
(2) 4	**4:2, 5:3, 6:3, 8:3**
(2) 7	7:3

To elaborate on this, associative classification and associating rule algorithms, such as CBA, find frequent items of size 1 (F_1) after scanning the training dataset and keeping all those which survived the minimum freq threshold. Then, disjoint items inside F_1 are appended to derive the candidate items of size 2 (C_2), and based on counting all items in C_2 they determine the frequent items of size 2 (F_2). Possible remaining candidate items of size 3 (C_2) are then obtained by merging disjoint items of F_2, counting them again, and so on. On the other hand, the proposed rule discovery method of eDRI relaxes this entire candidate generation function and only utilises items size one F1 without the need for: (1) joining disjoint items; (2)

passing over the dataset multiple times; and (3) undergoing an exhaustive counting process. Rather, the algorithm employs on the fly item frequency update procedures that are invoked whenever a rule is built and its data instances removed, which saves too many unnecessary costs associated with training time. It should also be noted another difference between associative algorithms, which discover the complete rules set and then adopt different pruning procedures, and Covering algorithms, which employ the rule-by-rule induction strategy.

The eDRI algorithm stores all surviving strong one items (F_1), items that are associated with a frequency greater than freq threshold, and seeks the item among them that when added to a class achieves the largest rule_ strength (item with the lowest error regardless of the class label). This item and class will be made an initial rule with the item placed in the rule's body and the class in the rule's consequent. This process of adding other item(s) to the rule is straightforward for eDRI since strong <item, class> of size one obtained from the initial scan are already ranked based on their frequency with the class labels in a descending manner. The candidate rule's data is located by eDRI, which continues adding items to the current rule's body until the rule either derives a zero error or one above the Rule_Strength threshold parameter. For the second scenario, i.e for any rule that cannot reach zero error, the algorithm checks whether its accuracy is greater than the Rule_Strength threshold. If the rule passes the Rule_Strength threshold, it will be created, otherwise being ignored.

Noticeable differences between the eDRI algorithm and most of the current Covering algorithms, such as PRISM and its successors, are as follows:

a) In eDRI, no item is added to the rule's body unless it verifies the minimum frequency requirement. Otherwise, the item is ignored and will not form part of any rule.

b) In eDRI, there is a possibility of creating rules that have accuracy lower than 100%, whereas PRISM only allows generation of perfect rules.

c) eDRI does not need to revisit the training data to amend item frequencies, rather it employs an efficient data structure procedure that guarantees dynamic update, thus saving computing resources.

In eDRI, whenever a rule is generated, the following applies

1. The rule's data samples in the training dataset are discarded.
2. Before building the next in line rule, eDRI updates item frequencies that have appeared in the removed data samples by decreasing

them to reflect the rule's training data samples deletion, without the need to scan the data again.

The proposed algorithm continues creating rules until no more items with a sufficient frequency can be found. At this point, eDRI moves to the remaining unclassified instances in the training data, and chooses the class that is linked with the more frequent data as a default class rule. In case the class labels in the remaining unclassified instances have equal frequency, eDRI will choose the largest frequency class in the initial training dataset. If class labels in the initial training dataset also have equal frequency, then eDRI selects the class that appeared with more generated rules. The pseudocode of the learning phase of eDRI is provided in Algorithm 3.2.

```
Input: Training set D, Rule_Strength, Minimum Frequency (freq)
thresholds
Output: Set of Rules (R)

Phase (1) Inducing the rules

1.  R ←θ
2.  r1 ←θ
3.  D′ ←D
4.  for each (atti, class) in D ′ do
5.    compute p(ci = i| atti)/| D ′ | and p(ci = i| atti)/ p(atti);
6.    remove any atti such that p(ci = i| atti)/| D ′ | <= freq
7.    append the largest (atti, class) accuracy to the current rule, ri ←atti
8.    if p [(ci = i| atti)/ p(atti)] < = Rule_Strength {
9.      D′ ←(D′ −(¬(p(ci = i | atti)))
10.     repeat steps 6-9 (keep adding single attribute values to the cur-
        rent rule in production)
11.     until ri >= Rule_Strength
12.                         } //if
13. R ←ri // adding the rule into the classifier
14. D′ ←(D −D′) // setting the temporary data after generating the rule
15. D ←D// setting up the new training data after removing the de-
    rived rule's data (after rj is generated)
16. ri←θ // re-initialising the current rule to empty for the next itera-
    tion
```

17. repeat steps 3-14 and stop when *D* is empty or no $p(c_i = i \mid att_i) >= freq$
18. } // for loop

Algorithm 3.2 eDRI learning phase

The rule discovery procedure keeps item ranks dynamic since item frequency with the class is continuously updated whenever a rule is derived. This dynamism provides a distinguishing advantage to the eDRI algorithm in determining items that become weak during construction of the classifier without needing to again look them up in the training dataset. By doing so, the search space of candidate items is minimised and should provide smaller in size classifiers. In fact, the proposed algorithm develops a pruning procedure that reduces over-fitting and results in rules with larger data coverage than PRISM (add a reference when this is proved/shown; maybe the experimental section of the thesis where this is covered). Indeed, the over-fitting is reduced for two reasons

1) There is a limit on the rule growth: a stopping condition that limits this growth is adopted.
2) Items with limited statistical significance (weak items) are discarded early, and the algorithm continuously cuts down item search space whenever a rule is discovered because of the dynamic update of the items frequency.

   To clarify, PRISM keeps adding items to the rule's body regardless of the number of data examples that might be covered by the rule. The focus of most Covering algorithms is maximising the rule's accuracy, even when the derived rule covers only one data example. This obviously may over-fit the training data and generates large number of specific rules that only cover very limited parts of the training dataset. For example, a simple run of the PRISM Covering algorithm over the "Weather.nominal" (only 14 data samples) dataset from the UCI repository revealed two rules covering a single data example each. In other words, 33.33% of the PRISM classifier's (two out of six rules) cover just two data examples (add a reference where this is actually reported). This result, if limited, shows how classic covering methods over-fit the training data without providing a red flag to stop rules during the learning process. In the case of the eDRI algorithm, these two rules are basically discarded since they do not hold enough data representation.

```
Input: Classifier (CL) and test dataset (Ts)
Output: One error rate
counter_hitting ←0
for each test data tsm in Ts {
   for each rule rn in CL {
      if (tsm = rn body)
         temp_rule ←rn }
   for each rule temp_rule {
      class_counter ++;
      tsm (class) ←Max(class_counter) (class) }
tsm (class) ←default class rule
If (tsm (class) == rn (class))
counter_hitting ←Counter_hitting + 1
}
Error ← Counter_hitting? / | Ts |
print Accuracy
```

$$Error \leftarrow \frac{Counter\_hitting?}{|\,Ts\,|}$$

Algorithm 3.3 Predicting test cases procedure of eDRI

Since the PRISM algorithm creates only rules with 100% accuracy, there is no rule preference procedure. Meanwhile, the eDRI classifier contains rules with good accuracy, though not necessarily 100%, and thus the algorithm utilises a rule sorting procedure to differentiate among results. The rule's strength and frequency are the main criteria employed to sort rules. When two or more rules have the same strength, eDRI uses a rule's frequency as a tie-breaker. Finally, whenever two or more rules have same strengths and frequencies, the algorithm prefers rules with a lower number of items in their body. This is because these rules are general rules with wider training representation that often hold multiple specific rules.

3.2.3 Test Data Prediction

The eDRI classifier consists of two rule levels

a) Level 1: Rules with zero error: Primary rules (higher rank)

b) Level 2: Rules with good strength: Secondary rules (lower rank)

Whenever a test data sample is about to be classified, eDRI goes over the rules in the classifier in a top down fashion starting with the primary rules. The first rule with items identical to the test data classifies it. In other words, if the rule's body is contained with the test sample, then this rule's class is assigned to the test data. If no rule in the primary rules matches the test data, then eDRI moves into the lower rank rules and evaluates whether any of its rules is able to assign a class to the test data sample. If one of the secondary rules in level 2 matches the test case attribute values, it will classify it. A more complicated scenario occurs when multiple rules are able to classify the test case. When this happens, eDRI groups them with respect to the class label into groups. It then counts the number of rules per group and the class label that belongs to the group with the largest number of rules is used to classify the test case. Finally, when no rule in the primary or secondary rules sets is able to classify the test case, eDRI invokes the default class rule (the largest frequency class in the training dataset).

This class prediction procedure limits the use of the default class rule, and decrease the number of misclassifications. In addition, eDRI also minimises the chances of arbitrary choices and biased decisions favouring a specific rule. Rather, eDRI employs the best first and combining with the group of rule prediction approach.

3.3 Example of the Proposed eDRI Algorithm and PRISM

This section provides an example to distinguish the learning process the typical covering algorithms PRISM and the proposed eDRI algorithm. In particular, we show how the rules are generated and the classifier building process is conducted alongside predicting test data cases. The dataset displayed in Table 3.3 is used for the purpose of comparison, assuming that the frequency threshold is set to 3 and the Rule_Strength to 80% for presentation purpose only. First, the eDRI algorithm calculates the frequency of items linked with each <item, class>. For presentation purposes, and because typical Covering algorithms work per class during data processing, eDRI starts with the minority class (No) and its items' frequencies are provided in Table 3.4. The largest frequent items are "Outlook=sunny" and """Windy=true", i.e. 3. The eDRI algorithm selects "Outlook=sunny" because it has higher accuracy, separating this rule's data samples as shown in Table 3.5, and forms the following first rule:

RULE (1) If "Outlook=sunny" then NO (3/5)

All items highlighted in red in Table 3.4 have failed to pass the frequency threshold; therefore they are ignored in the rule production of class "NO". This means that the eDRI algorithm has substantially minimised the search space for the rules by keeping only three candidate items (the ones not highlighted red in Table 3.4). However, PRISM keeps these candidate items, aiming to seek for specific rules with low data coverage.

The frequency of items in the data subset related to "Outlook=sunny" is computed to find out whether the current rule's strength can be improved, which is summarised in Table 3.6. It is clear that the highest rule's expected accuracy occurs when "Humidity=high" is added to the current rule, so eDRI appends this item to Rule (1) and the rule's strength becomes 100%. Therefore, eDRI generated Rule (1), removed all training samples connected with it (the red text of Table 3.3), and updated the frequency for the remaining candidate items of class "NO" as shown in Table 3.4 (Column 3). It should be noted that Table 3.5 is provided for presentation purposes although the eDRI algorithm does not need to scan such training examples but PRISM does. Based on the updated frequency of the items after Rule (1) is devised, item "Windy=true" becomes weak since its frequency has been decremented from "3" to "2". Therefore no more rules can be produced for the current class "NO", and the eDRI algorithm proceeds to the next class "YES".

RULE (1) If "Outlook=sunny and Humidity=high" then NO (3/3)

Table 3.3: Sample training dataset from (Witten and Frank, 2005)

outlook	temperature	humidity	windy	play	covering rule
sunny	hot	high	FALSE	no	R1
sunny	hot	high	TRUE	no	R1
overcast	hot	high	FALSE	yes	R2
rainy	mild	high	FALSE	yes	default
rainy	cool	normal	FALSE	yes	R3
rainy	cool	normal	TRUE	no	default
overcast	cool	normal	TRUE	yes	R2
sunny	mild	high	FALSE	no	R1
sunny	cool	normal	FALSE	yes	R3
rainy	mild	normal	FALSE	yes	R3
sunny	mild	normal	TRUE	yes	default
overcast	mild	high	TRUE	yes	R2
overcast	hot	normal	FALSE	yes	R2
rainy	mild	high	TRUE	no	default

Table 3.7 shows the candidate items associated with class "YES" along with their frequency and accuracy. The highest rule strength is associated with item "Outlook=overcast," its accuracy being 100%, and thus the following rule is formed:

RULE (2) If "Outlook=overcast "then YES (4/4)

After Rule (2) is generated, all data samples linked with it in Table 3.3 are discarded (the rows highlighted green), and the frequency of remaining candidate items of Table 3.7 are updated as shown in columns 4-5 of the same table. After the generation of Rule (2), the following three candidate items are pruned for having insufficient frequency as shown in Table 3.7: "Temperature= cool", "Humidity= high", and "Windy= true". This pruning, represented by constant item frequency updates whenever a rule gets generated, is indeed useful. The search space has reduced by 37.5% by discarding three out of eight candidate items after Rule (2) is derived. The algorithm then continues searching for rules linked with class "YES" from the remaining unclassified training data. Item "Humidity=normal" is then appended with the expected rule's accuracy of 80% as follow:

RULE (3) If "Humidity=normal" then YES (4/5)

Since the current rule has not yet reached maximum accuracy, its data samples are separated as shown in Table 3.8. Table 3.9 displays the remaining candidate item frequencies obtained from Table 3.8. The highest accuracy item, "Windy=false," is appended to Rule (3), achieving 100% accuracy, as shown below

Table 3.4: Candidate items linked with class NO

	During Rule 1 Generation		**After generating Rule (1)**		**Item Status**
Candidate Item	Original Frequency in the training data	Rule's Expected Accuracy (strength)	Freq. status		Item status
Outlook=sunny	3	60%			Picked largest expected accuracy
Outlook= rainy	2				Initially Ignored (fails freq. test)
Temperature= hot	2				Initially Ignored (fails freq. test)
Temperature= mild	2				Initially Ignored (fails freq. test)
Temperature= cool		25%			Ignored(fails freq. test)

Humidity= high	4	57.14%			Ignored after Rule(1) production (fails freq. test)
Humidity= normal	1				Initially Ignored (fails freq. test)
Windy= true	3	50%	2		Ignored after Rule(1) production (fails freq. test)
Windy= false	2				Initially Ignored (fails freq. test)

RULE (3) If "Humidity=normal and Windy=false" then YES (3/3)

Therefore, Rule (3) gets produced and all its connected data samples are removed (yellow rows of Table 1). The remaining items became weak because they have frequencies lower than 3 and therefore the algorithm stops. The four instances remaining in the training dataset are then covered by the default rule, which is the largest frequency class in the remaining uncovered data samples. In this case, we have a tie between the "NO" and the "YES" class labels, so the choice will be based on the largest frequency class in the training dataset.

Table 3.5: Data samples associated with item "Outlook=sunny"

outlook	temperature	humidity	windy	play
sunny	hot	high	FALSE	no
sunny	hot	high	TRUE	no
sunny	mild	high	FALSE	no
sunny	cool	normal	FALSE	yes
sunny	mild	normal	TRUE	yes

Table 3.6: Candidate items linked with item "Outlook=sunny" Then NO in the training dataset with their frequencies

	None	
Candidate Item	**Frequency**	**Rule's Expected Accuracy**
Temperature= hot	2	100%
Temperature= mild	1	50%
Humidity= high	3	100%
Windy= true	2	66.67%
Windy= false	2	50%

This results in the class "YES" default rule being built. Thus, the eDRI algorithm builds a classifier with following four possible rules:

RULE (1) If "Outlook=sunny and Humidity=high" then NO (3/3)

RULE (2) If "Outlook=overcast" then YES (4/4)

RULE (3) If "Humidity=normal and Windy=false" then YES (3/3)

Otherwise "YES"

In the above example, eDRI was able to induce three possible rules and a default class whereas PRISM produced six possible rules and a default class as shown below. In fact, most of the rules induced by PRISM cover very limited data examples (1 or 2 rows) whereas the eDRI algorithm rules cover at least three data examples. This makes a difference in practice, especially in large datasets or datasets with high dimensionality, as we will see in the experimental study carried out (Chapter 4). The fact that the eDRI algorithm derived a number of rules 42.85% lower than PRISM from a dataset of just 14 examples is evidence in itself of the power of the new pruning proposed which ensures the imposition of dynamic learning.

Table 3.7: Candidate items linked with class YES

	During Rule 1 Generation		**After generating Rule (1)**		**Item Status**
Candidate Item	**Original Frequency in the training data**	**Rule's Expected Accuracy (strength)**	**Freq. status**		**Item status**
Outlook=overcast	4	100%			
Outlook= sunny	2				Initially Ignored (fails freq. test)
Outlook= rainy	3	60%	3	60%	
Temperature= hot	2				Initially Ignored (fails freq. test)
Temperature= mild	4	66.67%	3	60%	
Temperature= cool	3	75%	2		

Humidity= high	3	42.85%	1		Ignored after Rule(2) production (fails freq. test)
Humidity= normal	6	85.71%	4	80%	
Windy= true	3	50%	1		Ignored after Rule(2) production (fails freq. test)
Windy= false	6	75%	4	66.67%	

Table 3.8: Training data samples linked with "Humidity=normal"

outlook	temperature	humidity	windy	play
rainy	cool	normal	FALSE	yes
rainy	cool	normal	TRUE	no
sunny	cool	normal	FALSE	yes
rainy	mild	normal	FALSE	yes
sunny	mild	normal	TRUE	yes

Table 3.9: Candidate items linked with item "Outlook=sunny" Then NO in the training dataset with their frequencies

	None	
Candidate Item	**Frequency**	**Rule's Expected Accuracy (strength)**
Outlook=rainy	3	66.67%
Outlook=sunny	2	
Temperature= cool	3	66.67%
Temperature= mild	2	
Windy= true	2	
Windy= false	3	100%

Prism rules

----

1) If outlook = overcast then yes
2) If humidity = normal and windy = FALSE then yes

3) If temperature = mild and humidity = normal then yes
4) If outlook = rainy and windy = FALSE then yes
5) If outlook = sunny and humidity = high then no
6) If outlook = rainy and windy = TRUE then no

Otherwise "YES".

3.4 Distinguishing Features of eDRI

In Covering classification, most of the different rule induction algorithms developed are based on on PRISM and IREP. The vast majority of Covering algorithms have been divised to deal with over-fitting or to enhance the processing of data during the induction of rules. Hereunder, the primary distinctive differences between eDRI and these existing induction algorithms are listed:

- PRISM and other Covering algorithms learn one rule then amend the training dataset, learn the second rule and again amend the training dataset, then continue learning in the same fashion. Therefore, each rule derived necessitates a full scan over the remaining training instances, which is an expensive computational data processing approach. On the other hand, eDRI adopts a special data structure and dynamic procedure and whenever a rule is generated the remaining item frequencies are updated on the fly without the necessity of training data scans, thereby saving processing time and computing resources.
- eDRI employs a data processing procedure that limits the search space of items by discarding items with limited statistical significance during the process of building rules. Other Covering algorithms keep all items without paying attention to their training data representation for the sake of inducing highly accurate rules, yet with low data coverage which may increase the chance of over-fitted classifiers.
- The classification procedure employed by eDRI minimises random rule selection and ensures that more than one rule is used for assigning the class of test data. This results in less questionable prediction decisions and reduces the dependency on a single rule, unlike most Covering methods that employ just one rule for class prediction.
- eDRI splits rules discovered in the classifier into two levels. In the secondary level, eDRI keeps rules with some error in case rules in the primary level were unable to predict a test data. This is unlike classic Covering approaches, like PRISM, which only derive rules

with zero error without considering their training data coverage. Overall, eDRI allows the production of rules with small errors in order to reduce over-fitting of the training data set and the number of discovered rules in the classifier.

- eDRI employs an implicit pruning procedure that is naturally integrated with the data processing phase. In pruning, eDRI prevents items from participating in a rule early and ignores rules that do not meet the Rule_Strength requirement. This significantly cuts down both the search space of rules as well as the expected number of derived rules. The result is a more efficient induction phase and smaller model that facilitates their management and usage by users as efficient tools for making decisions.
- eDRI adopts a data structure from the distributed association rule that prevents level-wise searching in rule induction algorithms such as IREP and RIPPER. This data structure holds all information about items, classes, and their occurrences in the training dataset and is used for rule and item pruning by eDRI.

3.5 Chapter Summary

Covering is a promising classification approach that has attracted researchers due to its simplicity. This chapter dealt with major shortcomings associated with classic Covering algorithms. Specifically, we investigated the problem of generating rules with limited data coverage by PRISM and discarding good quality rules covering larger training data portions, leading to massive classifiers. A new dynamic Covering algorithm called Enhanced Dynamic Rule Induction (eDRI) has been proposed and built into the WEKA environment (Chapter 4) to deal with these deficiencies. The eDRI algorithm minimises the search space for rules by cutting down items with low data representation and stopping learning when any rule meets the rule's strength threshold. Moreover, eDRI implements a multiple prediction rule method that reduces doubtful prediction decisions and thus minimises random rule selection. In addition, eDRI reduces the need for repetitive data scans during the data processing phase, resulting in a more efficient rule induction method. These improvements guarantee smaller sized classifiers that do not over-fit the training dataset and maintain their predictive performance.

Classifiers derived by the proposed eDRI algorithm are of benefit to managers, particularly when making key business decisions since they can serve as a rich knowledge base inside a decision making tool. This is due to the classifiers being easy to understand, having high predictive accuracy, and being controllable as well as robust (flexible to be amended without drastic change). Moreover, eDRI can be employed as a prediction module embedded inside a decision support system in various domain applications that necessitate both simplicity of the outcome and good accuracy, such as medical diagnoses systems or anti-phishing models in computer security (Chapter 5). Since general practitioners usually have busy clinics with tightly allocated time per patient, they favour decision support systems with a concise set of knowledge similar to the eDRI classifiers. Regarding computer security, eDRI is planned to be embedded by researchers in Huddersfield University as an anti-phishing tool based on the case study conducted in Chapter 5. This will raise awareness of public employees about phishing because of eDRI's outcome simplicity, which can be understood by different users.

In Chapter 4, we present the detailed implementation of eDRI, its main thresholds used, and the graphical user interface of the algorithm with which end-users interact. We also highlight the evaluation measures used to evaluate the strengths and weaknesses of eDRI, along with a full description of the large number of datasets used in the experimentations. Furthermore, experiments using datasets from the UCI repository have been conducted utilising a number of Covering, induction, and decision tree algorithms to measure eDRI's performance with reference to various evaluation measures in classification.

Chapter 3

Implementation and Evaluation of the Enhanced Dynamic Rule Induction Algorithm

4.1 Introduction

In this chapter, we discuss the implementation of the new proposed eDRI algorithm and explain the different thresholds it utilises for the dynamic learning process. Since the proposed algorithm has been implemented in the WEKA environment (see Section 4.2), its primary Graphical User Interface (GUI) is also briefly highlighted in Section 4.2 (Hall, et al., 2009). After this, the integration of eDRI within the classifier package of WEKA under "Classifier.Rules" is covered and a discussion of the input parameters used is provided. Furthermore, a few GUIs of eDRI within WEKA are shown, in particular the different processes to go through until the predictive model is generated. Section 4.3 explains the evaluation measures (one-error rate, confusion matrix, number of rules derived, etc.) used to produce the results and test the performance of the proposed eDRI algorithm and other known predictive models in classification for their comparison (Diebold & Mariano, 1995; Powers, 2011).

This Chapter also includes a number of experiments on datasets published at the UCI data repository (see Section 4.4) (Lichman, 2013). In particular, twenty different datasets that belong to various applications (credit card scoring, medicine, politics, and agriculture among others) are utilised. The chosen datasets have different number of variables, variable types, and data examples. Section 4.4 highlights the main characteristics of the datasets used in the experiments. The experimental evaluation validate eDRI's advantages, particularly in predicting test data as well as limiting data processing time. Later, in Chapter 6, a further specific evaluation and comparison of eDRI with a wide range of algorithms on a phishing problem is reported. The choice of the phishing application was based on the fact that it is a vital threat that necessitates not only accurate detection but also useful knowledge for decision making.

4.2 eDRI Implementation

WEKA stands for Waikato Environment for Knowledge Analysis and is an open source tool based on the Java platform that contains implementations for different DM methods including filtering, classification, clustering, evaluation, and visualisation among others. This tool was designed and implemented at New Zealand's Waikato University to assist students, researchers, and academic staff in conducting quantitative research. Experiments on the datasets using the considered classification algorithms were conducted using WEKA. The proposed eDRI algorithm has implemented in Java and integrated into the WEKA environment.

Since WEKA is built with Java programming language, it organises the different filtering, learning, and visualisation methods in packages. A package can be seen as a container that holds and manages related Java classes and has its own subdirectories (Bouckaert et al. 2008). In WEKA versions 3.7.2 and upward, information other than classes can be held inside a package such as the functionalities of the JAR file, meta data, source codes, and other related documentations of the classes. These enable users to have better management of which methods can be stored in separate packages, and therefore can use what they actually need. By default, WEKA keeps packages and their related information in $WEKA_HOME, which is located in the user's home directory (i.e. user.home/WEKAfiles), although it can be using an environment variable (more information can be found in Bouckaert et al. 2013).

The main class of the proposed algorithm, "EDRI," has been integrated within the Rules package inside WEKA. This Rules package contains implementation of a few rule-based algorithms such as JRIP, PRISM, ZeroRule, Decision Table, PART, and OneRule. Since the proposed algorithm generates classifiers with an "If-Then" rules format, it was more relevant to include it inside "Classifiers.Rules. eDRI uses two input parameters (shown in the appendix), "minFrequency" and "minRuleStrength," in addition to an optional selection to use a default rule during the classification process. The proposed algorithm can be accessed from WEKA Explorer or the Command Line platforms for data processing.

The minimum frequency threshold is used to reduce the number of items (variable values) that may participate in creating the rules. By using this threshold, the eDRI algorithm will be able to discard early items that their data representation is below the *minFrequency* value (weak items) and thus to keep statistically significant items (strong items). Moreover, the minRuleStrength threshold is employed by the proposed algorithm to produce not only rules with 100% positive data examples, as in typical

Covering algorithms, but also rules that may cover a few negative data examples as long as their computed strengths are greater than or equal to the *minRuleStrength* value. This threshold sustains the predictive performance of eDRI by only keeping rules that are effective yet have a large data coverage in the training dataset. Both the minFrequency and minRuleStrength must be inputted as decimals by the end user in (see Appendix) before eDRI starts data processing.

The minFrequency plays critical role in the data processing phase. If this threshold has been set to very high value, i.e. 1, then more specific rules that cover limited training data instances will be generated. On the other hand, if it has been set to very low value, i.e. <1%, then possibly more rules are derived. We have conducted warming up experiments to evaluate this threshold and found out that values between 1% to 4% balance out the number of rules generated as well as data coverage. Unfortunately, no ideal value for the minFrequency that works for any dataset simply because datasets have different characteristics. A recently published distributed associative classification (see Thabtah, et al., (2015)) showed that minFrequency has no ideal value and setting it remains a challenging question in both association rule and classification domains. Therefore, the requirements of both the decision maker and application play important role in setting the minFrequency. Nevertheless, and based on our experiments as well as Thabtah, et al., 2015, we recommend 1%-4% to be assigned for the minFrequency.

Finally, as mentioned before, there is an option within the eDRI interface, "addDefaultRule", which enables the user to utilise a default class rule in cases when all derived rules are unable to classify an unseen data case. The user has this option so a default class rule can be utilised when no other rules are available to classify a test case, and thus the number of random classification will be minimised.

Three types of dataset format are possible to be received in eDRI; comma separated formats in spreadsheet (CSV), text formats that are comma separated (TXT) similar to WEKA's ARFF file format, and an external database that can be connected with WEKA using ODCB or JDCB connectivity objects. Furthermore, all evaluation measures and cross validation testing method are offered in eDRI because we have taken advantages of WEKA's other classes and packages (Kohavi, 1995). Mathematical and explanatory details on the evaluation measures are given in Section 4.3.

For model evaluations, eDRI can use separate test datasets, training datasets, or cross validations with different variations in terms of the number of folds or split percentages. In all experiments of eDRI, reported in Section 4.4, the testing method used was tenfold cross validation. Appendix A provides a detailed example of data processing using eDRI on one dataset from UCI for further graphical exposition.

4.3 Evaluation Measures

A number of measures related to predictive models in classification that we used are discussed in this Section. In particular, we describe the testing method, cross validation, common predictive model evaluation criteria such as accuracy, one-error rate, and confusion matrix related measures such as false positive, false negative, true positive, true negative, etc. Moreover, methods related to computing resources utilised through the data processing of the learning algorithms such as processing time and numbers of data examples are also covered. We also present two new methods for rule evaluation in this Section: average rule length (ARL) and weighted average rule length (WARL).

4.3.1 Cross Validation

One of the major issues for minimising biased results in predictive models within classification is the use of cross validation (Kohavi, 1995). Cross validation ensures that the models derived from data processing are evaluated on data examples that have not been seen during the training phase. In ten-fold cross validation, the dataset is randomly partitioned into ten different parts. The model is trained on nine parts, and then examines the remaining "hold out" part to derive an average error rate (Equation 4.1). Error rate denotes the proportion of training examples in the hold out part that the model has misclassified from the total number of data examples. In ten-fold cross validation, the process of partitioning the training data is repeated ten times and the derived errors are averaged.

One of the challenges in cross validation is random partitioning of the data examples, which may cause a class imbalance problem. In binary classification, this results in a majority class that has much larger data instances than the minority class in the input data. Therefore, to guarantee each class is represented in each partition, so the produced predictive model can be generalised, cross validation must be stratified. Stratification ensures distribution of all available classes in the training data into each partition when the random shuffling occurs.

4.3.2 Confusion Matrix

For DM and ML predictive models, a matrix called the error table, or the confusion matrix, has been developed (Powers, 2011; Fawcett, 2006). The confusion matrix is typically used to evaluate the performance of predictive models with respect to the different metrics that are primarily related to the predictive power of the models. Predicted and actual values of the class labels, as shown in Table 4.1, are then displayed in the confusion matrix. Each row in the table represents data examples of the actual class, while each column represents the data examples of the predicted class.

In measuring the performance of the models in classification, a test case is assigned a predicted class by the model. If the test case class is identical to the predicted class, this counts for a correct classification; otherwise it is considered a misclassification. For a test dataset with M cases, the error rate (Equation 4.1) denotes the proportion of the misclassified cases from M. On the other hand, classification accuracy (Equation 4.2) denotes the proportion of correct classifications from M.

Relevancy measures such as recall (Equation 4.3) and precision (Equation 4.4) are also useful for evaluating predictive models (Rocchio, 1971). In Information Retrieval, recall is the proportion of relevant data examples that are retrieved whereas precision is the proportion of retrieved data examples that are relevant. In classification problems, and for a test data with N (= *15*) cases, ten cases have "yes" class and N-10 (5) cases have "no" for the actual classes. For the cases with actual class=yes, assume the predictive model was able to predict "yes" for seven (7) of them, and the remaining three (3) cases were misclassified as "no". For the cases with actual class = no, four (4) were predicted correctly and one (1) wrongly. Thus, the model's recall is 70%, while its precision is 87.5%. Relevancy measures are usually used when the dataset is imbalanced with respect to the class label since accuracy becomes a biased measure toward the majority class label. Notice that there are measures more relevant to multi-label classification problems and therefore these have been omitted from the experimental section (Tsoumakas, et al., 2010).

Table 4.1: Confusion matrix for ASD diagnosis problem

	Predicted Class	
Actual Class	**YES**	**NO**
YES	True Positive (TP)	False Negative (FN)
NO	False Positive (FP)	True Negative (TN)

$$One\_error(\%) = 1 - Accuracy \tag{4.1}$$

$$Accuracy(\%) = \frac{|TP + TN|}{|TP + TN| + |FP + FN|} \tag{4.2}$$

$$\text{Re}\,call(\%) = \frac{|TP|}{|TP + FN|} \tag{4.3}$$

$$\text{Pr}\,ecision(\%) = \frac{|TP|}{|TP + FP|} \tag{4.4}$$

4.3.3 Computing Resources Evaluation Measures

Measuring the processing time taken for the predictive model to be constructed is a useful indicator of the efficiency of the data processing phase. Moreover, the time taken to predict data cases can be measured as part of the model's efficiency. In the WEKA environment, we recorded the time taken to build the predictive model in milliseconds (ms) so we can compare eDRI processing time performance with other classification algorithms. Moreover, we have also computed how many instances are necessary to build the predictive model for the eDRI and PRISM Covering algorithms, and implemented it as a Java class in the WEKA version. This class contains codes that compute repetitive data scans, and can be seen as a performance indicator of the proposed algorithm when compared with other Covering algorithms.

4.3.4 New Rules Evaluation Measures

Initially, we have used the number of rules produced by the classification algorithms as a measure of goodness, that is the trade-off between the number of rules generated and the predictive accuracy of the classifier. We have sought classifiers with a smaller number of rules since users are able to manage them in a more effective manner. Furthermore, new rule evaluation measures based on the number of variables inside the rule and how many instances a rule may cover are proposed here. These measures reflect the relationship between the rule length in the classifier and the number of instances the rule has covered. The first rule measure is called the average rule length (ARL) and the second is known as the weighted average rule length (WARL) as per the following expressions (4.5) and 4.6, respectively:

$$\text{ARL} = \sum (\text{\# of rules of length } n * \text{ rule length n})/Total\ \#\ of\ rules \tag{4.5}$$

$$\text{WARL} = \sum_{i=0}^{n} ri's\ length * \frac{\#\ of\ data\ example\ covered\ by\ ri}{size\ of\ the\ training\ dataset} \tag{4.6}$$

To elaborate on these two measures, consider for instance the following two rules:

- Rule 1: (if a=windy and b = mild then yes)
- Rule 2: (if a=sunny then no)

Rule 1 has a body length of two attribute values, and let us assumes that it covers three data instances. On the other hand, rule 2 has one attribute value in its body and let us assumes a coverage of 97 data instances. Based on this information, one can calculate ARL and WARL as follows:

- ARL = (1+2)/2 = 1.5
- WARL = 2 * 3/100 + 1 * 97/100 = 1.03

It seems that the rules set has ARL=1.5 which indicates that the average rule length is 1.5 variables. On the other hand, the weighted averahe rule length of 1.03 indicates that correlation between the rules set and the number of covered data instacnes in the training dataset.

4.4 UCI Data Experimental Settings

In this section, different datasets from the University of California Irvine (UCI) repository used to evaluate the proposed eDRI algorithm performance are described (Lichman, 2013). In particular, twenty datasets along with their specifications as displayed in Table 4.2 have been used. All numerical attributes of the chosen datasets have been discretised, and missing values were replaced using ReplaceFilter in WEKA (Hall, et al., 2009).

Table 4.2: The UCI datasets characteristics

dataset	# of attributes	# of instances	# of classes	Continuous attributes	Missing values
Arrhythmia	280	452	16	Yes	Yes
Balance-scale	5	625	3	Yes	No
Cleve	12	690	2	Yes	Yes
Credit-g	21	1000	2	Yes	No
Cylinder-bands	40	540	2	Yes	Yes
Dermatology	35	366	6	Yes	Yes
Pima_diabetes	9	768	2	Yes	No
Hayes-roth-test	5	28	4	Yes	No
Hayes-roth-train	5	132	4	Yes	No
Hepatitis	20	155	2	Yes	Yes
Hypothyroid	30	3772	4	Yes	Yes

Ionosphere	35	351	2	Yes	No
Liver-disorders	7	345	2	Yes	No
Lung-cancer	57	32	2	No	Yes
Lymph	19	148	4	Yes	No
Mushroom	23	8124	2	No	Yes
Sick	30	3772	5	Yes	Yes
Tae	6	151	3	Yes	No
Tic-tac-toe	10	958	2	No	No
Waveform	41	5000	3	Yes	No

For fair comparison, datasets with different sizes have been chosen. Stratified ten-fold cross validation method, as explained in Section 4.3.1, has been used for testing all considered classification algorithms. This method is widely used in ML and DM communities to produce average error rates of the classifiers (Witten and Frank, 2005).

In order to exhibit eDRI's performance with respect to different measures when contrasted with a wide range of ML algorithms that adopt a variety of learning strategies, a number of highly competitive classification algorithms that generate rule-based models are required to be utilised to conduct the experiments (Quinlan, 1993; Cendrowska, 1987; Cohen, 1995; Holt, 1993; Witten & Frank, 2005; Freund & Schapire, 1997; Liu, et al., 1998; Gaines & Compton, 1995). The selection of rule-based classifier algorithms was based on four main factors:

a) Produce rules are in the format of "If-Then", so they are similar to the proposed eDRI algorithm;
b) They adopt different rule discovery mechanisms;
c) All are known algorithms that have been evaluated by previous researchers in the DM and ML communities;
d) They are all implemented in the recent developer version of WEKA 3.7.

. The above criteria resulted in the following classification algorithms that generate rule-based models to be selected for the experiments: decision trees C4.5, PRISM, RIPPER, OneRule, Conjunctive Rule, Boosting (AdaBoost is the exception with respect to criterion a) above), CBA, and Ridor were chosen.

It should be noted that the majority of PRISM successors such as P-PRISM and N-PRISM focus on treating noise datasets and reducing computing resources (parallelism) as described earlier in Chapter two. Thus, their rule learning strategy is identical to PRISM learning startegy, and therefore PRISM WEKA's version was selected for comparison. Moreover, REP, IREP, IREP++ employ similar learning strategies to the RIPPER one, and considering that RIPPER is more advanced because of its optimisation procedure it was therefore the selected one to be part of the experimentations.

All experiments have been run on a Core i5 computer with a 3.1 GHz processor and 8.0 GB RAM. The minimum frequency and rule strength thresholds for eDRI have been set in all experiments to 1% and 50%, respectively. Minimum frequency threshold plays a critical role in controlling the number of rules that may be generated, and therefore following recent published studies (Thabtah, et al., 2015; Abdelhamid, et al., 2014) we set it to 1%. On the other hand, rule strength has a low impact so was set to 50%. These values normally balance between the classifier's size and the classification accuracy performance measure as shown by other scholars (Thabtah, et al., 2016; Abdelhamid et al., 2014; Li et al., 2001).

We utilised a number of evaluation measures to show the benefits and negatives of the proposed algorithm when compared with other classification algorithms in DM. Precisely, the below measures have been used to evaluate eDRI:

- One error rate (%)
- The classifier size (Number of rules derived)
- The time in ms taken to discover and extract the rules

4.5 UCI Data Results

The eDRI algorithm was tested against twenty generic classification datasets that belong to different applications at the UCI data repository. The choice of the dataset was based on different data features such as number of data examples, number of class values, available number of variables in the dataset, attribute types, data application domain type, and noise tolerance such as missing values, etc. Overall, datasets with small, medium, and large dimensionalities have been selected as shown in Table 4.2 provided before.

Figure 4.1 depicts the average one-error rate generated by the considered algorithms on the twenty UCI datasets. The CBA algorithm was unable to produce rules from twelve datasets (often large dimensionality ones) due to a combinatorial explosion problem (running out of memory), so we omitted its average error results and will later show its results on the eight datasets it produced rules. Figure 4.1 clearly shows that the eDRI algorithm outperformed the considered algorithms on average, with an average error,lower than PRISM, C4.5, PART, Ridor, OneRule, ConjunctiveRule, RIPPER, and AdaBoost by 9.75%, 4.41%, 3.67%, 4.19%, 13.46%, 11.34%, 2.32%, and 8.49%, respectively. Its dynamic rule production by live updating the frequency and rank values of each candidate rules ensures fair and high quality rules. These play a major role in decreasing the error rate of the classifiers produced by the eDRI algorithm when compared with other rule learning algorithms. Furthermore, the proposed eDRI algorithm eliminates any item with insufficient data representation from becoming part of any rule's body. This is done by evaluating each item's frequency against the frequency threshold while testing each rule-strength during the learning phase against the minimum rule-strength threshold. These two early pruning tests ensure that only high quality rules are kept for predicting test data cases.

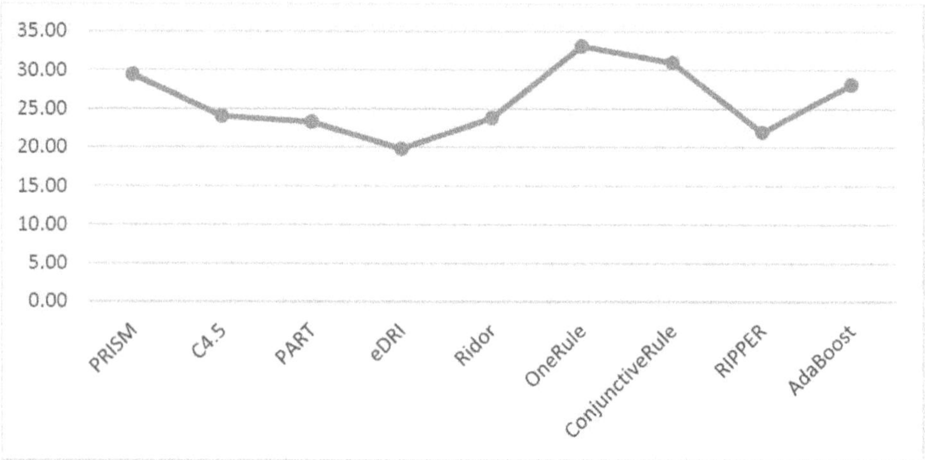

Figure 4.1 Error rate generated by the considered classification algorithms against the 20 datasets

In order to further evaluate the predictive power of the proposed eDRI algorithm, the error rate per dataset for all considered algorithms are provided in Table 4.3. The figures clearly show a consistent domination for the eDRI algorithm when compared to the remaining algorithms. In particular, eDRI won-lost-tie record against PRISM, C4.5, PART, Ridor, OneRule, ConjunctiveRule, RIPPER, and AdaBoost are 18-2-0, 9-9-2, 11-8-1, 11-8-1, 17-3-0, 17-3-0, 9-9-2 and 15-5-0, respectively. It seems that RIPPER, Ridor, PART, C4.5, and the proposed eDRI algorithm are more predictive than the remaining rule-based classifiers, as it is obvious from both the average error and the per-dataset errors generated. This makes OneRule, ConjunctiveRule, and RPSIM the least applicable algorithms due to the low predictive models derived. Models derived from datasets such as "Arrhythmia", "Balance-Scale" and "Ionosphere" make this especially clear.

Table 4.3: The considered algorithms error rate for the UCI datasets

Dataset	Algorithms								
	Ridor	C4.5	PART	RIP-PER	PRISM	OneRule	Conjuncti-veRule	Proposed eDRI	Ada-Boost
Arrhythmia	38.94	38.49	42.69	29.64	61.50	42.47	46.68	39.38	44.46
Balance-scale	20.48	35.52	22.72	20.96	36.32	43.68	37.28	16.16	27.68
Cleve	57.42	13.62	14.20	48.18	21.03	46.53	27.06	16.10	16.83
Credit-g	28.10	27.20	30.70	28.30	36.20	33.90	30.5	28.08	30.50
Cylinder-bands	31.11	42.22	40.74	34.81	44.80	50.37	34.44	26.22	27.40
Dermatology	6.83	6.01	5.19	13.11	15.56	50.27	49.72	8.46	49.72
Pima_diabetes	25.00	26.17	26.56	23.95	38.92	28.51	31.25	27.08	25.65
Hayes-roth-test	46.42	50.00	50.00	53.57	57.14	50.00	53.57	14.28	39.28
Hayes-roth- train	18.93	27.27	25.75	14.39	31.05	56.06	57.57	20.45	56.81
Hepatitis	21.29	18.70	19.35	21.93	22.57	20.00	21.29	20.64	17.41
Hypothyroid	0.55	6.73	7.26	0.66	8.77	3.76	2.94	6.70	6.78
Ionosphere	11.96	13.39	12.82	10.25	13.95	19.08	18.51	13.39	9.11
Liver-disorders	36.81	38.84	37.68	35.36	44.34	44.92	42.31	35.36	33.91
Lung-cancer	37.50	21.87	25.00	21.87	41.62	12.50	18.75	25.00	21.87
Lymph	14.86	20.27	19.59	22.29	24.31	25.00	27.02	20.27	25.67
Mushroom	0.00	0.00	0.00	0.00	0.00	1.47	11.32	0.49	3.80

Sick	1.82	1.74	2.22	1.77	1.95	3.60	2.28	1.80	2.81
Tae	48.34	50.33	52.98	58.94	45.02	53.64	34.43	43.04	62.25
Tic-tac-toe	8.03	15.44	5.74	2.19	3.54	30.06	30.06	8.03	27.45
Waveform	22.28	27.22	25.20	20.80	39.42	46.28	42.68	22.00	33.36

The advantage of eDRI becomes apparent whenever a rule is inserted into the classifier and its covered data are discarded, thereby causing an immediate amendment on the remaining potential rule rank. This is because the classifier constructed contains rules with little to no redundancy, which ensures that:

a) No training example is used multiple times in inducing rules;

b) Each item has its real time frequency rather than the initial frequency computed from the training dataset;

c) Rule rank, which is the primary measure for rule strength, is constantly updated. This safeguards the rule discovery phase since insignificant rules are removed during learning despite some of them having a high rank at the first scan.

The development of rules linked with constant frequency and rank values have contributed to the decrease of the one-error rate in the classifiers derived by eDRI. As a matter of fact, the algorithm ensures each rule is derived from the remaining instances in the training data after removing instances associated with the already generated rules. This only allows rules that are constantly fit statistically to participate in the classifier. These rules are the ones utilised later on during the class prediction step.

The average number of rules in the classifiers produced by C4.5, PART, PRISM, and eDRI is depicted in Figure 4.2. We omit the rules generated from OneRule and ConjunctiveRule since both are single-rule learners. Moreover, RIPPER uses excessive pruning based on a growing set and a pruning set, besides an optimisation procedure, thus a very limited number of rules are generated. Therefore, for fair comparison, we omitted the classifier of RIPPER. Later we show RIPPER classifiers for the phishing problem since that problem necessitates small yet useful numbers of rules. Figure 4.1 illustrates that PRISM and C4.5 produce classifiers with a larger number of rules than the remaining algorithms, which make it hard for end-users to manage and understand. On the other hand, eDRI and Ridor were able to produce, on average, an acceptable size of predictive model. Despite Ridor's ability to get smaller classifiers than eDRI, the predictive power of eDRI is still higher and may be seen as a definite advantage. Ridor "must" perform extensive rule pruning, and therefore usually ends up with few rules.

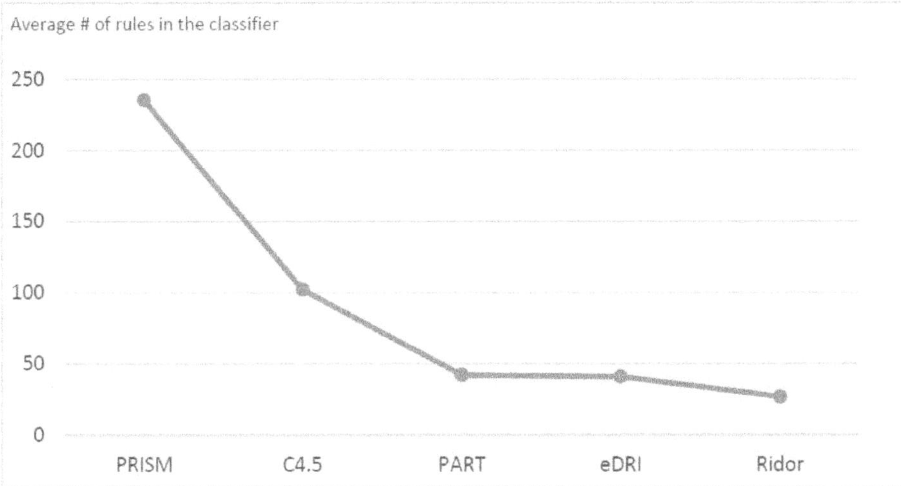

Figure 4.2 Average # of rules generated by the selected rule based classification algorithms against the 20 datasets

PRISM generates larger classifiers than the rest of the considered algorithms and on most of the datasets considered, which is due to the fact that PRISM prefers rules with low data coverage (specific rules). However, eDRI on average generates a lower number of rules in the classifier than C4.5, PART, and PRISM. We believe that this rules reduction is attributed to two main reasons:

1) Each rule covers a large number of training instances because of incremental data processing.
2) The new learning strategy employed by eDRI allows a rule to cover more training instances.

The mechanism of rule learning in eDRI contributed to a decrease in the final classifier because when each rule is inserted into the classifier eDRI reduces the search space of remaining potential rules by only storing those that are linked with acceptable frequency and rank values. Other algorithms, such as PART and C4.5, build trees based on information gained and repeatedly evaluate all variables in the training data. When a variable is linked with many possible values, this may result in many branches, each of which can be linked to a leaf node and thus making a rule. During the building of the classifier, these algorithms may generate a larger number of rules despite employing backward and forward pruning. On the other hand, eDRI induces and evaluates each rule in parallel until the dataset is empty or no item with sufficient data is present. This substantially decreases the available number of candidate items for the next possible rules.

We also looked at the time required to build the predictive models in average milliseconds (ms) (Figure 4.3) and per dataset (Table 4.4). According to Table 4.4, eDRI consistently built classifiers in an efficient manner. In fact, eDRI was able to outperform all tested algorithms with respect to average time taken to construct the predictive models, and for each individual dataset. Figure 4.3 demonstrates that eDRI achieved 7, 3, 9, 49, and 4 less milliseconds in inducing the rules than the PRISM, C4.5, PART, Ridor, and AdaBoost algorithms respectively. These figures are clearer when the number of input variables are large, as exhibited in cases such as the "Wavefor" dataset. Therefore, we expect that when the dataset consists of a larger number of attributes the gap in time taken to find the rules between eDRI and the remaining algorithm will increase. Minimisation in time during the process of rule discovery in eDRI is attributed to the fact that this algorithm has combined rule discovery and building the classifier into a single phase. In eDRI, no separate classifier building procedure is included as in traditional rule induction algorithms such as PART, Adaboost, Ridor, and C4.5.

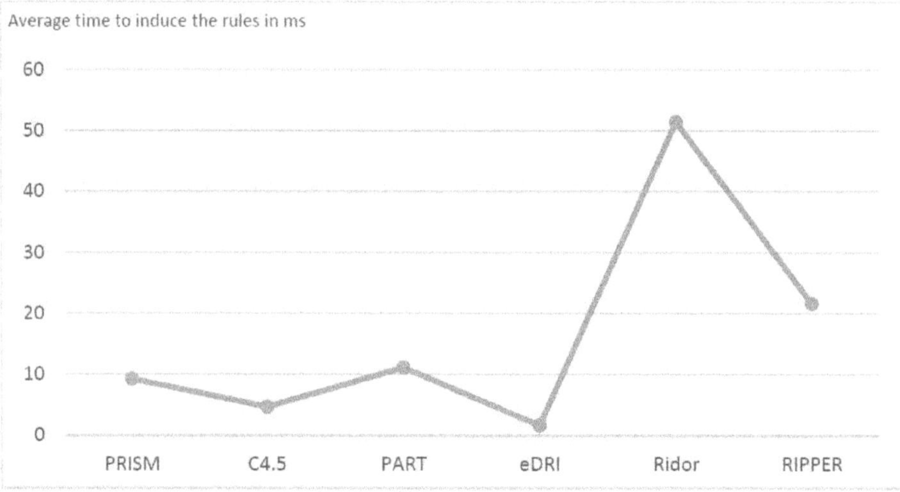

Figure 4.3 Average time in ms generated by the considered classification algorithms against the 20 datasets

Table 4.4: The considered algorithms time in ms to generate the rules from the 20 UCI datasets

Time to induce the rules in ms							
dataset	PRISM	C4.5	PART	eDRI	Ridor	RIPPER	AdaBoost
Arrhythmia	22	17	38	8	688	72	32
Balance-scale	3	10	2	0	2	3	0
Cleve	3	2	5	0	2	2	0
Credit-g	6	3	5	2	5	9	2
Cylinder-bands	5	0	11	2	3	5	2
Dermatology	0	2	2	0	8	3	2
Pima_diabetes	2	3	3	2	2	2	2
Hayes-roth-test	0	0	0	0	0	0	0
Hayes-roth-train	0	0	0	0	0	0	0
Hepatitis	2	0	0	0	0	0	0
Hypothyroid	6	9	24	5	17	11	5
Ionosphere	2	3	5	0	2	3	2
Liver-disorders	1	1	3	1	0	2	2
Lung-cancer	1	1	1	1	0	0	0
Lymph	0	0	0	0	0	0	0
Mushroom	2	5	4	2	14	29	11
Sick	7	8	8	2	6	22	12
Tae	3	4	1	2	0	0	1
Tic-tac-toe	1	2	2	1	2	3	2
Waveform	119	23	108	5	278	266	35

We have included a new rule-based associative classification algorithm called CBA to see how eDRI performs in predictive accuracy and rule generation with this competitive family of algorithms (Liu, et al., 1998). As indicated earlier, CBA crashed when the number of attributes increased, so no results for CBA on twelve out of the twenty datasets could be generated on the WEKA platform.

Figure 4.4 displays the error rate derived respectively by all considered algorithms on the eight datasets that CBA was able to mine clearly shows that CBA results are competitive when compared to those obtained by eDRI, PART, C4.5, and PRISM algorithms, which support previous research findings in AC research literature (Thabtah, et al., 2015; Abdelhamid net al., 2014). To be more precise, CBA was able to produce, on average, higher predictive classifiers than C4.5, PART, and PRISM on the eight datasets while eDRI slightly outperformed CBA in terms of error rate on the same datasets. Nevertheless, classical AC algorithms have shown a significant drawback in mining datasets with moderate to large numbers of attributes, which can be seen as a drawback of this family of algorithms because it limits their use in commercial applications.

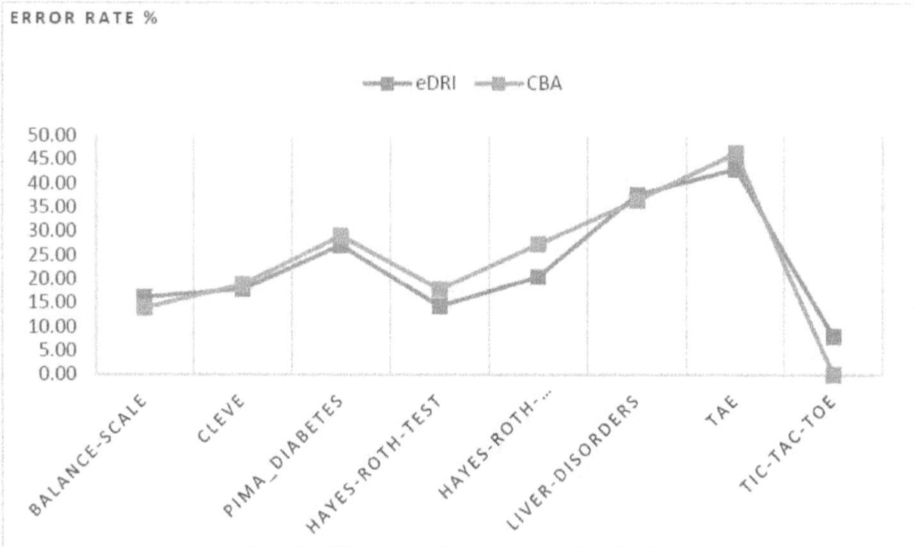

Figure 4.4 Error rate generated by CBA and the considered classification algorithms against the 8 least # of attributes datasets

The fact that CBA failed to mine twelve datasets is evidence of its exhausting rule discovery and evaluation phases. Relaxing such exhaustive mining by eDRI is a bonus that not only produces higher quality classifiers in terms of predictive accuracy but also smaller in size ones. It is clear from the experimentation that CBA produced larger classifiers than eDRI on seven out of the eight datasets mined. In fact, CBA derived 38 rules more than eDRI on average for the eight datasets with a limited number of attributes it was able to mine. We can only deduce that this average difference will increase when mining datasets with higher dimensionality.

Finally, the runtime in milliseconds for the classification algorithms on these eight datasets has been recorded in Table 4.5. The figures clearly point out that CBA is the slowest algorithm to construct classifiers, which seems a new trend in AC mining approaches. This has been attributed to two factors: CBA "must" discover all correlations in the training dataset using the candidate generation function; and it employs a tense rule evaluation step that requires for each training instance a full scan over the complete set of candidate rules before selecting the classifier. These two steps unfortunately consume a tremendous amount of time and are a substantial deficiency in CBA. According to Table 4.5, CBA has to spend 299 ms more on average than eDRI to build the classifiers, which places it in last place among the considered algorithms in terms of efficiency.

Table. 4.5 Runtime in ms generated by CBA and the considered classification algorithms against the 8 least # of attributes datasets

PRISM	C4.5	PART	eDRI	CBA
3	10	2	0	7.7
3	2	5	0	1720
2	3	3	2	540
0	0	0	0	0
0	0	0	0	5.7
1	1	3	1	33.5
3	4	1	2	9.7
1	2	2	1	88

We investigated the search space used by PRISM and eDRI to determine the major differences between both algorithms in constructing the classifier since both are the only classic Covering algorithms. The number of data samples (rows) scanned by both algorithms in determining item frequencies and accuracy while building the rules were recorded. Table 4.6 illustrates the findings. For most of the datasets considered, PRISM had to scan more data samples to derive the final classifier except for the "Labor" dataset. One example is the "Segment Challenge" dataset which consists of 19 attributes and 1500 instances where PRISM has to pass over 7,147,542 rows while eDRI only used 938,263 rows. On average for the datasets we examined, PRISM and eDRI had to scan 1,075,551 and 203,350 times, respectively. This is clear evidence to support that the proposed eDRI algorithm substantially reduces the search space for rules in its classifier building process. The fact that eDRI on average reduces the search space by around five times less than PRISM is significant since it stops building rules when they have an acceptable error rate, rule strength, according to the user.

Table 4.6: # of scanned data samples during building the classifier by PRISM and eDRI on the UCI datasets

Dataset	PRISM	eDRI
Contact lenses	1503	168

Weather	1160	196
Labor	13796	19507
Glass	325951	58613
Iris	20613	15265
Diabetes	1979939	418680
Segment Challenge	7,147,542	938,263
Zoo	72,852	21,011
Tic-Tac	509,652	120,526
Pima	682,505	358,451

4.6 Chapter Summary

In this chapter, the details of eRDI implementation were provided and its integration in the WEKA environment has been discussed, explaining the required parameters in the graphical user interface. In addition, we described the different evaluation measures used, the mathematical notations required, the experimental settings, and the algorithms used in the experimental comparison conducted.

A number of experiments have been carried out in the WEKA environment using twenty datasets from the UCI data repository to measure the success and failure of the proposed eDRI algorithm. We utilised a number of evaluation metrics such as predictive accuracy, number of rules, classifier content, time taken to construct the predictive models in ms, etc. A wide range of predictive models that normally produce classifiers with rule formats have been contrasted with eDRI with respect to the aforementioned evaluation metrics. Experimental results revealed that eDRI is an efficient DM algorithm that constantly derives highly competitive predictive models when compared to those derived with Ridor, PRISM, PART, C4.5, and AdaBoost. Specifically, eDRI was able to outperform the considered algorithms regarding average error rate and for most datasets. Furthermore, the predictive models derived by eDRI were often moderate in size and smaller than the rest of the contrasted algorithms except for Ridor. Nevertheless, Ridor continuously consumed larger quantities of time in producing the predictive models, making this algorithm the least applicable when processing large amounts of data. The dynamic adjustment in item and rule frequencies and rank contributed to the efficient data processing and higher predictive model performance of eDRI. The results of the experiments can be summarised as follows:

1) eDRI outperformed the considered algorithms in regard to their predictive power for most datasets and, on average, for the entire datasets.
2) The classifiers produced by eDRI were smaller than those of PRISM, C4.5, and PART, which makes it applicable to applications that necessitate concise sets of knowledge for decision making. In Chapter 6, we show the effectiveness of eDRI in detecting phishing websites.
3) The efficiency of eDRI in processing small, medium, and large datasets have been measured and it has been shown its superiority over the considered algorithms.
4) The new dynamic Covering learning in eDRI scaled well when contrasted with the high predictive learning in associated classification approaches such as CBA. Results showed superiority not only in classification accuracy of the models generated, but also in the time taken to build and size of the models.

In the next chapter, the problem of phishing within the context of website security is investigated. We then apply eDRI on a real dataset related to phishing detection in order to measure the strengths and weaknesses of eDRI as an anti-phishing tool, comparing its performance with a wider selection of ML algorithms (both rule and none rule based).

Chapter 5

Applicability of eDRI on Cyber Security Applications: Phishing Detection

5.1 Introduction

As the largest computer network, the Internet is considered a critical platform for business success and expansion as most commercial trades are being conducted online. With the majority of businesses competing in a global market, they seek to maximise revenue by increasing user accessibility and promotion of products and services through the web-based and mobile platforms. People not only use the internet for socialisation and knowledge but purchasing goods, pay bills, and transfer money, and consequently the internet has become a necessity for many individuals. Since the explosion of mobile commerce applications access to the World Wide Web has become a daily essential requirement, with people performing regular financial transactions regardless of their geographical location (Ronald et al., 2007; Aburrous et al., 2010b).

With the advanced development of computer hardware, especially computer networks and cloud technology services, online and mobile commerce have significantly increased in the last few years (Abdelhamid & Thabtah, 2014). Indeed, the number of customers who perform online purchase transactions has dramatically increased and large monetary values are daily exchanged through electronic means, such as private payment gateways, that are usually verified by secure socket layer (SSL) (Sheng et al., 2010). Despite the convenience associated with online transactions from both user and business prospectives, an online threat has emerged: phishing.

Phishing involves creating a well designed website that replicates an existing authentic business website in order to deceive users and obtain illegally their credentials and/or login information (Abdelhamid, 2015), so as for phishers to get access to legitimate users' financial information (Afroz, et al., 2011). Unfortunately, the consequences of phishing are fatal because affected legitimate users become vulnerable to identity theft and information breach and no longer trust online commerce and electronic banking (Nguyen, et al., 2015). Phishing typically occurs via email sent to users, from apparent trustworthy sources, urging them to adjust their log-in information by clicking/following a hyper link within such email (Khadi & Shinde, 2014).

In 2011, Gartner Group published a report (McCall, 2011) that revealed an annual loss over $2.8 billion as a result of phishing activities occurring within the United States. For this reason, an international body that aims to minimise online threats including pharming, spoofing, phishing and malware, the Anti-Phishing Work Group (APWG), was created (Jevans, 2003). APWG periodically disseminates reports for the online community on recent cyber-attacks, with a recent report stating the rapid increase of phishing websites to 17,000 in the month of December 2014 alone (Aaron & Manning, 2014). In the same report, the total number of monthly phishing activities reported in the fourth quarter of 2014 was more than 197,000 websites, and increase of 18% compared with the previous quarter of the same year, with the United States continued to be the top country hosting phishing websites.

It seems imperative that users, as well as businesses, adopt renewable anti-phishing tools or strategies to reduce phishing activities and protect themselves from their potential negative impacts. This is important because phishing attacks are constantly changing and new deceptions are emerging all the time. Anti-phishing solutions adopting DM (ML) are shown to be more practical and effective in combating phishing because they work automatically and are capable of revealing concealed knowledge that online users are not aware of, especially with respect to the relationship among website features and phishing activities. This hidden knowledge, when combined with human experience, can result in an effective shield for protecting users from phishing (Abdelhamid, 2015).

In this Chapter, we investigate the phishing problem and define it in a ML context. We then discuss common, traditional, strategies in addition to computerised techniques developed to combat phishing. More importantly, this Chapter thoroughly investigates ML anti-phishing techniques and critically analyses both their benefits and drawbacks.

Specifically, we evaluate the eDRI model proposed in Chapter three on the hard problem of phishing detection in order to exhibit its applicability as an anti-phishing tool. Using real data, we compare eDRI with other common ML algorithms and in reference to different evaluation metrics.

5.2 Phishing Background

Phishing comes from the word "fishing," in which the phisher throws a bait and awaits for potential users to take a bite. Phishing is not recent as an online risk, with its origin rooted in a social engineering method using telephones known as "phone phreaking" (Rader & Rahman, 2015). It was during the 1990s period when the internet community started to grow that phishing was originally observed as an online threat, especially in the United States (Basnet, et al., 2008).

Figure 5.1. Example of an early phishing attack (Blogonlymyemail.com)

According to McFredies (2016), the first phishing incident was noticed in the mid-1990s when phishers attempted to obtain registered online users' account information from the internet provider America Online (AOL). Phishers during this era frequently utilised instant messages (IM) in AOL chat rooms or emails to reach users so that they would reveal their passwords (Figure 5.1 illustrates an example of an early phishing attack), which were sunsequently used by phishers to leverage the victims' accounts and begin emailing spam to other online users. Obvioulsy, phishers realised that they could further trick victims if the IMs and emails requested them to update their billing information. With this realisation, the attackers expanded their aim and using the same electronic means (IMs and emails) attempted to access other financial information from victims such as social security numbers, addresses, credit card information, etc.

One of the most common beliefs that ordinary users have about phishing websites is that grammatical errors and typos are typical within these websites (Rader & Rahman, 2015). This has misled users into believing that a website without grammatical errors must be trustworthy, which has proven not necessarily be true. Phishers in today's cyber world become more innovative and work systematically in groups sometimes even orchestrating phishing campaigns motivated by potential financial gains. In fact, phishers are continuously changing their spoofing methods based on the counterpart security measures taken by organizations, and recently they have directed their attention to the mobile commerce platform. This means that the profile of phishers has changed from egocentric purposes into more organized and serious cybercrime that keeps evolving, which makes it hard to detect.

5.2.1 Phishing Process

Phishing attacks are initiated through an email sent to potential users. Other ways a phisher may start an attack include Instant Messaging, online blogs and forums, short message services, peer to peer file sharing services, and social media websites (Abdelhamid, et al., 2014). We can summarise the phishing lifecycle as follows (see Figure 5.2):

1) A link is sent using one of the aforementioned channels to potential victims.
2) When clicked, the link will redirect potential victims to a malicious website.
3) Users become vulnerable as they try to login using their credentials on the malicious website.
4) Login credentials are then transferred to a server, or a key logger is installed into the user's computing device.
5) The phisher can then utilise the credentials to perform additional cybercrimes.

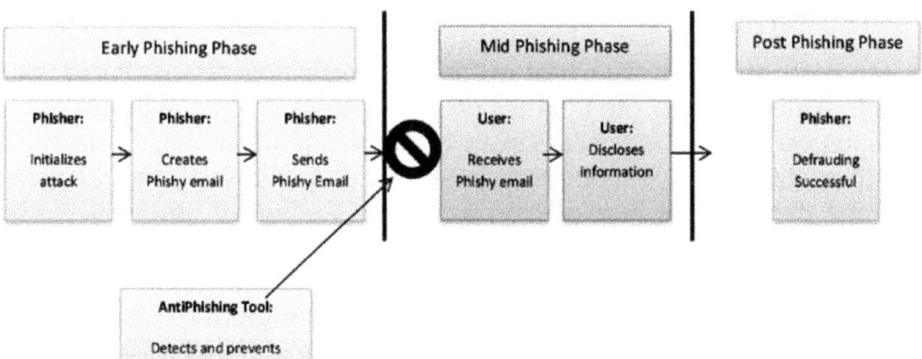

Figure 5.2. Phishing life cycle (Abdelhamid et al., 2014)

5.2.2 Phishing as a Classification Problem

Generally speaking, websites can be classified by hand-crafted methods based on certain features such as URL length, prefix_suffix, domain, sub_domain, etc. Initially, scholars in the area of online security (Aburrous et al., 2008; 2010b) developed different knowledge bases using their experience and expertise to distinguish phishing from legitimate websites. Rcently, there have been studies and proposals aiming at deriving intelligent rules to detect the fine line between legitimate and phishing websites using statistical analysis (Qabajeh and Thabtah, 2015; Mohammad et al., 2014b; Abdelhamid et al., 2014). For instance, Aburrous et al. (2010a) and Mohammad et al. (2014b) defined a number of hand crafted rules based on various website features using simple statistical analysis on websites (instances) collected from different sources including Phishtank and Yahoo directory (Phishtank, 2011). More advanced decision rules have been developed in Abdelhamid et al. (2014) in which the authors used further statistical analysis on a larger phishing dataset collected from varying sources.

ML and DM have proved to be powerful data analysis tools in many application domains such as medical diagnosis, market basket analysis, weather forecasting and events processing, to cite some (Abdelhamid & Thabtah, 2014). This is due to the fact that ML and DM techniques usually reveal concealed meaningful information from large datasets so they can be utilised in management decisions related to development, planning, and risk management. Generally speaking, ML and DM can be seen as an automated and intelligent tool embedded within management information systems to guide decision making processes in both business and scientific domains. Common tasks or problems that ML and DM handle are clustering, association rule discovery, regression analysis, classification, pattern recognition, time series analysis, trends analysis, and multi-label learning (Witten & Frank, 2005).

One of the frequent task of ML is the forecasting of a target variable within datasets based on other available variables (Witten & Frank, 2005). This forecasting process occurs in an automated manner using a classification model, normally named the classifier, that is derived from a labelled training dataset. The goal of the classifier is to "guess" the value of the target variable in unseen data, referred to as the test dataset, as accurately as possible. This task description falls under the umbrella of supervised learning and is known as classification. Abdelhamid and Thabtah (2014) defined classification as the ability to "accurately" predict class attributes for a test instance using a predictive model derived from a training dataset.

Since the problem of website phishing involves automatic categorisation of websites into a predefined set of class values (legitimate, suspicious, phishy) based on a number of available features (variables) then this problem can be considered a classification problem. To be more specific, the training dataset will consist of a set of predefined features and the class attribute and instances are basically the websites' feature values. These instances can be extracted from different sources such as Phishtank and online directories. The aim will be to build an anti-phishing classifier that can predict the type of website based on hidden knowledge discovered from the training set features during the data processing phase. Usually the goodness of the classifier is measured using accuracy, which primarily relies on the correlations of the features and the class (Thabtah et al., 2016a). Figure 5.3 shows phishing as a classification problem from the ML prospective. As discussed before, phishing websites are dynamics and consequently it can modelled as a dynamic supervised learning classification problem. Therefore, an effective anti-phishing classifier should be adaptable to any new features observed in order to handle and manage the dynamic nature of the problem.

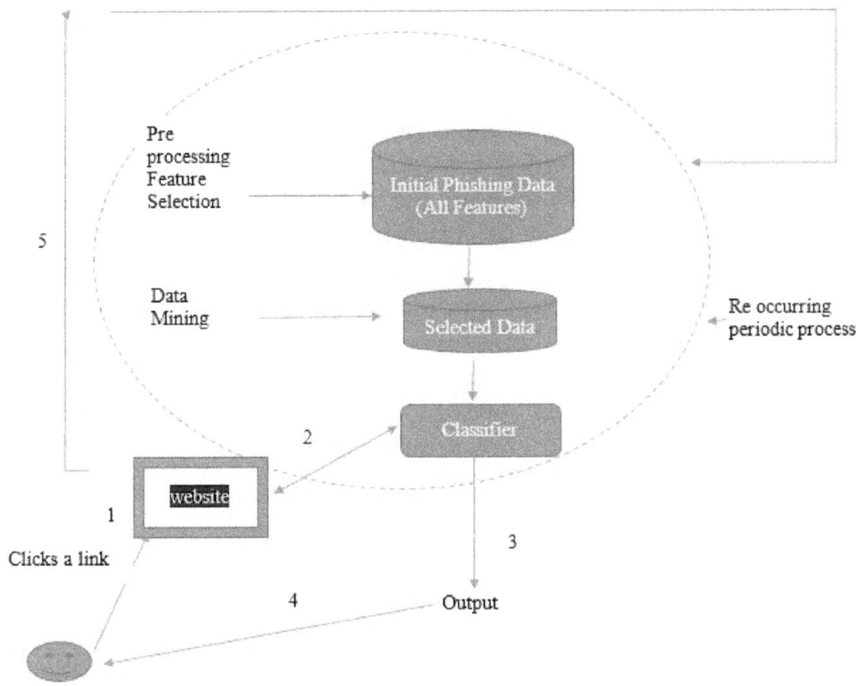

Figure 5.3. Phishing as a classification process

5.3 Common Traditional Anti-Phishing Methods

Since phishing causes serious breaching of user confidentiality, as well as organizations including government agencies, there have been different methods proposed to combat it. These approaches can be categorized into three main groups:

- Education and legal
- Computerized using human-crafted methods
- Intelligent ML methods

In this section, we examine the literature on phishing and critically analyse different techniques based on the above categories. The focus, however, will be on the intelligent anti-phishing solutions because that is the aim of this PhD dissertation but also because it is our belief this will be the way forward in shielding the web from phishing threats as promising recent results confirm (Thabtah et al. 2016b; Nguyen et al. 2015; Lee et al. 2015; Mohammad et al. 2014a; Jameel and George 2013; Khadi and Shinde 2013; Sheng et al. 2010).

5.3.1 Legal Anti-phishing Legislations

Governments have been slow in responding and opposing started to oppose phishing. California State in the US was the first to issue anti-phishing legislation in 2005 (InformationWeek, 2016). This legislation stated that it is unlawful to use any electronic means such as websites, emails, or any other methods to ask or solicit information from online users by claiming ones self as a business without the authority of that business. Other US States such as Texas have also introduced new cybercrime legislations that include phishing, and in 2005 the General Assembly of Virginia added phishing attacks to their list of computer crimes (General Assembly of Virginia, 2005). These new laws empowered companies such as America Online to file lawsuits in Virginia against phishers in 2006 (Pike, 2006). However, most US States have not yet designated laws for phishing and usually prosecute phishers using other computing crime laws such as fraud.

At Federal level in US, lawmakers and congressional representatives have not passed anti-phishing legislation either. There were a few attempts between 2004 and 2006 following the Anti-Phishing Act of 2004 to pass specific bills incriminating phishing and instigating tougher prison sentences, but these bills were stopped at the committee level in Congress. Nevertheless, federal law enforcement can incriminate phishers using other laws that are related to identity theft and fraud such as "18 U.S.C. section 1028" (Pike, 2006). Businesses have also joined the Government in fighting phishing. For example, in 2005 Microsoft filed over 115 lawsuits in Washington's Western District Court accusing a single Internet user of utilising various deceptive methods to access some of the company's users' information (Pike, 2006). In mid-2006, the then president George W. Bush established a new cybercrime identity theft task force (Executive-Order-13402, 2006), with a single goal: reduce the risks of cybercrime, especially phishing.

The United Kingdom (UK) has followed the US by strengthening its legal system against severe cybercrimes, including fraud and identity theft. In 2006, the UK introduced the new Fraud Act, which increased prison sentences to up to ten-years for online fraud offences (BBC News, 2005). This same act prevents possession of a phishing website with the intent to deceive users and commit fraud. Further, Microsoft decided to collaborate with other law enforcement agencies outside US to bring justice to phishers. In doing so, the company signed an agreement with the Australian government to train law enforcement agents in preventing phishing (Government of Australia, 2011). Also, in 2010 Canada introduced an Anti-spam Act that incriminates cybercrime and that aims to protect Canadian online consumers and businesses when globally trading (ClickDimensions, 2014).

5.3.2 Educating Users: Simulated Training

One of the easier, yet helpful, policies to oppose cybercrimes is to educate users on the ways employed to access their information. Users awareness of the circumstances around phishing will undoubtedly hep minimise this risk or stop it as early as possible. Unfortunately, most users are unaware of how phishing attacks start or how to recognise visually an untruthful website and differentiate it from one that is trustworthy (Mohammad et al., 2015). Moreover, basic security indicators and anti-phishing software counterparts are still vague for many online shoppers (Qabajeh & Thabtah, 2015). All this contributes to phishing increase and, subsequently, motivate phishers to launch further attacks. A security survey conducted by Julie et al. (2007) revealed the lack of knowledge on cybercrimes, including phishing, held by online users, and that some security aware respondents to the survey were reluctant to use their financial information for payment purposes even within trustworthy websites.

There have been a number of studies on educating people as to the severity of phishing. For example, Arachchilage and Love (2013) investigated whether mobile games can be a helpful method for raising awareness of phishing attacks. The authors evaluated learning curves of users who played with a mobile game about phishing developed by Arachchilage and Cole (2011), and assessed whether an interactive mobile platform is effective in educating users in contrast to traditional security training. A comparison of user responsiveness to phishing has also been conducted using the developed mobile game, along with a website designed by APWG. The results showed that users who played the anti-phishing mobile game were able to spot non-genuine websites with a higher rate of accuracy than other users who only used the APWG website. A number of organizations and research studies (Arachchilage et al. (2007);

Ronald et al. 2007) have adopted training to warn users of phishing, which involves sending participants simulated malicious emails from a genuine source to evaluate their vulnerability to phishing. Embedded training is another way to measure users' vulnerability to phishing. This type of training often mimics primary daily business processes performed by employees by experimenting to measure a certain outcome (Arachchilage, et al., 2013). Based on a preliminary pilot study conducted on a limited number of students, Arachchilage et al. (2007) used embedded training methodology to measure phishing awareness at a University. The authors sent malicious emails from the administrator to participants without informing them of the training material content. These emails urged users to click on a link that would redirect them to a malicious website where they would input their login credentials. This aim was directed at identifying the number of users who would actually click on the link. During the experiment, the user was interrupted immediately when he clicked the link and was then provided with the training material.

5.3.3 User Experience: Anti-phishing Online Communities

One of the simplest approaches to reducing the impact of phishing on online users and organisations is to build an anti-phishing community to monitor recent phishing activities and provide news to the different stakeholders. Users' experiences are practical and based on real cases related to different types of phishing. Such efforts by users and organizations have resulted in new proactive online communities and data repositories. These accumulated and useful resources are of interest since they can be employed to study ways to make the Internet safer and free from phishing.

The Monitoring and Takedown (MaT) approach enables individuals who recognise phishing activity to report it via public anti-phishing communities including APWG, PhishTank, Millersmiles, and Symantec among others (Jevans, 2003; PhishTank, 2011; Bright, 2011; Nahorney, 2011). These anti-phishing communities allow users to report phishing content and warn other users and organisations as well. Users can also report phishing content to the Federal Trade Commission's Complaint Department, becoming directly part of the campaign toward combating phishing. Many reputable companies also have an internet fraud department that allows users to report any fraudulent or suspicious activity such as phishing. PhishTank was created in 2003 as a subsidiary of OpenDNS in order to provide the parent company, as well as the online community, with a phishing repository. This large collection of stored phishing websites has given computer security experts, users, researchers and business owners' extensive information about phishing attacks and the features of their associated emails and websites. Another example of a good use of user experiences is Cloudmark, which is an alerting-based anti-phishing method with user rating system (Cloudmark, 2002). When a user is visiting a website and experiencing any kind of threat, they can then rate that website to alert other online users. Finally, Web of Trust (WOT) is another example of an anti-phishing approach based on the user feedback rating model (Web of Trust, 2006).

5.3.4 Discussion on Non Intelligent Anti-phishing Solutions

Legislators in the US, UK and Canada, among others countries, have approved legislative bills that include serious jail sentences for incriminated phishers. This has been made clear in several high profile cases, especially in the US. Nevertheless, these legislative bills have not achieved a decrease of phishing attacks. On the contrary, phishing has now become more severe than ever and businesses as well as individual users have suffered from substantial financial losses as a result. One of the primary reasons for legal actions not to be as effective as expected in minimising phishing is due to the fact that often a phishing website has a short life span (normally about two days), which helps the phisher to disappear quickly once the fraud has been committed, making law enforcement difficult.

As previously mentioned, raising awareness of phishing risks and educating users has shown promising initial results (Ronald et al. 2007). Computer security scholars have adopted different ways to disseminate the seriousness phishing may cause to society, with Jevans (2003) and PhishTank (2011) using web-based material to teach novice users phishing fraud techniques; while others, such as Arachchilage et al. (2007), developing contextual and embedded trainings based on simulated phishing emails coming from genuine sources; or educational material on phishing based on mobile games in order to increase the motivation factor among (Arachchilage et al. 2009; 2013)..

Even though educating users may positively impact the global efforts of combating phishing, this approach demands high costs and requires users to be equipped with computer security knowledge. Large organizations and governments are periodically investing in the development of anti-phishing materials in both hard and soft forms as well as websites and mobile applications. However, since phishing techniques keep changing/evolving, small to medium enterprises might not have the resources large organisations have to enable them to invest in their users' education. Therefore, a large portion of the online community realistically cannot afford the continuous additional costs to keep updating current anti-phishing material. Furthermore, phishing techniques are becoming more sophisticated because of the group efforts of phishers who employ systematic attack strategies, which make it harder for even security experts and specialised law enforcement agents to keep their skills updated. This makes ordinary users vulnerable, even if they were equipped with basic knowledge about phishing. Thus, more advanced, cheaper and intelligent approaches are needed for their implementation both within educational and legislative solutions to further reduce phishing attacks. We have seen thoughtful attempts that evolved from user experiences, user ratings, and users' social networking (such as Phishtank, Cloudmark, and APWG among others helping novice users and enterprises avoid falling prey to phishing). Effectiveness of these user community based approaches relies mainly on the following factors: (1) User experience; (2) User knowledge; (3) User honesty; and (4) Accessibility and validity of the user community's website data. Unfortunately, these factors are difficult to measure and validate, thus relying on user experience and knowledge alone necessitates carful care and accuracy. We hypothesise that by "only" considering users' experience in judging a websites' legitimacy is not enough to combat phishing, although it can be a supporting approach to a more advanced intelligent solution based on ML/DM.

5.4 Computerised Anti-Phishing Techniques

There has been development of anti-spam software tools that can block suspicious emails, however, these programs constantly block a large number of genuine emails and classify them as junk emails (Mohammad, et al., 2015). Emails misclassified as spam are simply false positive instances. Thus, one of the ultimate goals of the computerised anti-phishing tool is to reduce false positives and increase true positives so users can be confident of their mailbox's filter results without having to manually check their junk email folder.

5.4.1 Databases (Blacklist and Whitelist)

A database driven approach to fight phishing, called blacklist, was developed by several research projects (Site Advisor, 2006; Sheng, et al., 2009; Google SB, 2010). This approach is based on using a predefined list containing domain names or URLs for websites that have been recognised as harmful. A blacklisted website may lose up to 95% of its usual traffic, which will hinder the website's revenue capacity and eventually profit (Aburouse et al., 2010b). This is the primary reason that web masters and web administrators give great attention to the problem of blacklisting. According to Mohammad et al. (2015), there are two types of blacklists in computer security:

- **Domain/URL Based**. These are real time URL lists that contain malicious domain names and normally look for spam URLs within the body of emails.
- **Internet Protocol Based.** These are real time URL or domain server blacklists that contain IP addresses who, in real-time, change their status. Often, mailbox providers, such as Yahoo for example, check domain server blacklists to evaluate whether the sending server (source) is run by someone who allows other users to send from their own source.

Users, businesses, or computer software enterprises can create blacklists. Whenever a website is about to be browsed, the browser checks the URL in the blacklist. If the URL exists in the blacklist, a certain action is taken to warn the user of the possibility of a security breach. Otherwise, no action will be taken as the website's URL is not recognized as harmful. Currently, there are a few hundred blacklists which are publically available, among which we can mention the ATLAS blacklist from Arbor Networks, BLADE Malicious URL Analysis, DGA list, CYMRU Bogon list, Scumware.org list, OpenPhish list, Google blacklist, and Microsoft blacklist (Return Path 2016). Since any user or small to large organisation can create blacklists, the currently public available blacklists have different levels of security effectiveness, particularly with respect to two factors:

1) Times the blacklist gets updated and its consistent availability.
2) Results quality with respect to accurate phishing detection rate.

Marketers, users, and businesses tend to use Google and Microsoft blacklists when compared with other publically available blacklists commonly use because of their lower false positive rates. A study by Sheng et al. (2009) analysing blacklists concluded that they contain on average 47% to 83% phishing websites.

Blacklists often are stored on servers, but can also be available locally in a computer machine as well (Mohammad et al., 2014b). Thus, the process of checking whether a URL is part of the blacklist is executed whenever a website is about to be visited by the user, in which case the server or local machine uses a particular search method to verify the process and derive an action. The blacklist usually gets updated periodically. For example, Microsoft blacklist is normally updated every nine hours to six days, whereas Google blacklist gets updated every twenty hours to twelve days (Mohammad et al., 2015). Hence, the time window needed to amend the blacklist by including new malicious URLs, or excluding a possible false positive

URLs, may allow phishers to launch and succeed in their phishing attacks. In other words, phishers have significant time to initiate a phishing attack before their websites get blocked This is an obvious limitation of using the blacklist approach in tracking false websites (Abdelhamid, et al., 2014). Another study by APWG revealed that over 75% of phishing domains have been genuinely serving legitimate websites and when blocked imply that several trustworthy websites will be added to the blacklist, which causes a drastic reduction in the website's revenue and hinder its reputation (Aaron & Manning 2014).

After the creation of blacklists, many automated anti-phishing tools normally used by software companies such as McAfee, Google, Microsoft, were proposed. For instance, The Anti-Phishing Explorer 9, McAfee Site Advisor, and Google Safe Base are three common anti-phishing tools based on the blacklist approach. Moreover, companies such as VeriSign developed anti-phishing internet crawlers that gather massive numbers of websites to identify clones in order to assist in differentiating between legitimate and phishing websites.

There have been some attempts to look into creating whitelists, i.e. legitimate URL databases, in contrast to blacklists (Chen & Guo, 2006). Unfortunately, since the majority of newly created websites are initially identified as "suspicious," this creates a burden on the whitelist approach. To overcome this issue, the websites expected to be visited by the user should exist in the whitelist. This is sometimes problematic in practise because of the large number of possible websites that a user might browse. The whitelist approach is simply impractical since "knowing" in advance what users might be browsing for might be different to those actually visited during the browsing process. Human decision is a dynamic process and often users change their mind and start browsing new websites that they initially never intended to.

One of the early developed whitelist was proposed by (Chen & Guo, 2006), which was based on users' browsing trusted websites. The whitelist monitors the user's login attempts and if a repeated login was successfully executed this method prompts the user to insert that website into the whitelist. One clear limitation of Chen and Guo's method is that it assumes that users are dealing with trustful websites, which unfortunately is not always case.

Phishzoo is another whitelist technique developed by Afroz and Greenstadt (2011). This technique constructs a website profile using a fuzzy hashing approach in which the website is represented by several criteria that differentiate one website from another including images, HTML source code, URL, and SSL certificate. Phishzoo works as follows:

1) When the user browses a new website, PhishZoo makes a specific profile for that website.
2) The new website's profile is contrasted with existing profiles in the PhishZoo whitelist.
    - If a full match is found, the newly browsed website is marked trustful.
    - If partly matching, then the website will not be added since it is suspicious
    - If no match is found but the SSL certificate is matched, PhishZoo will instantly amend the existing profile in the whitelist.
    - If no match is found, then a new profile will be created for the website in the whitelist.

Recently, Lee et al. (2015) investigated the personal security images whitelist approach and its impact on internet banking users' security. The authors utilised 482 users to conduct a pilot study on a simulated bank website. The results revealed that over 70% of the users during the simulated experiments had given their login credentials despite their personnal security image test not being performed. Results also revealed that novice users do not pay high levels of attention to the use of personal images in ebanking, which can be seen as a possible shortcoming for this anti-phishing approach.

5.4.2 Intelligent Anti-Phishing Techniques based on DM

Since phishing is a typical classification problem, ML and DM techniques seem appropriate for deriving knowledge from website features that can assist in minimising the problem. The key to success in developing automated anti-phishing classification systems is a website's feature. Since there are a tremendous number of features linked with a website, a necessary step to enhance the predictive system performance is to pre-process the set of features in order to pick up the "most" effective. Feature effectiveness can be measured using different computational intelligence methods such as information gain, correlation analysis, and chi-square among others (Quinlan, 1979; Hall, 1999; Liu & Setiono, 1995).

Once an initial features set is chosen, the intelligent algorithm can be applied on the selected features to come up with the predictive system. There are many ML and DM algorithms for classification that have been developed by scholars in the last two to three decades as covered in Chapter 2. Most of these algorithms use one of the following major classification approaches in deriving their predictive systems:

1) Decision trees (ID3, C4.5 and successors)(Quinlan, 1993).
2) Probabilistic models (Naïve Bayes, Bayesian Network and successors)(Duda & Hart, 1973; Friedman et al., 1997).
3) Rule-based classification
    a. Associative classification (AC)
        i. Classification based Association (CBA and successors) (Liu et al., 1998).
        ii. Classification based on Multiple Association (CMAR and successors) (Li et al., 2001).
        iii. Multiclass Classification-based Association (MCAR and successors) (Thabtah et al., 2005).
    b. Rule induction such as FOIL, RIPPER and successors (Quinlan, 1979; Cohen, 1995).
    c. Covering or greedy, such as PRISM (Cendrowska, 1987) and eDRI (Thabtah, et al., 2016).
4) Neural Networks (NN) methods and their successors (Grossberg, 1988).
5) Support Vector Machine (SVM) (Joachims, 1999)
6) Fuzzy Logic (FL) (Zadeh, 1965)
7) Boosting and paging methods, and their successors (Freund & Schapire, 1997).
8) Search methods such as Genetic Algorithms (GA) (Goldberg, 1989)

Among the aforementioned classification approaches, rule-based classification systems are more suitable as an anti-phishing tool because: (1) their proven merits in predicting target values in many domains, including medical diagnoses, stock market analysis, email classification, text categorization, etc; and (2) the content of rule-based classification systems is simply human knowledge that novice users can easily understand and apply when necessary. Often the knowledge is formed as "If-Then" rules in which the antecedent of the rule ("If" part) consists of the conjunction of attribute values (Feature values) and the consequent of the rule ("Then" part) containing the target attribute value (Website type).

The rest of this section critically analyses intelligent anti-phishing attempts based on ML. We show how these approaches derive a classification anti-phishing system along with their benefits and weaknesses.

5.4.2.1 *Decision Trees*

Fette et al. (2007) explored email phishing utilising the C4.5 decision tree classifier among other methods including Random Forest, SVM and Naïve Bayes. As a result, a new Random Forest method called "Phishing Identification by Learning on Features of Email Received" (PILFER) was developed. Experiments on a set of 860 phishy and 695 ham emails were conducted. Various features for distinguishing phishing emails identified included: IP URLs, time of space, HTML messages, number of connections inside the email, and JavaScript. The authors claim that PILFER can be improved towards grouping messages by joining all ten features discovered in the classifier apart from "Spam filter output".

5.4.2.2 *Rule Induction*

Mohammad et al. (2014b) investigated a number of rule induction algorithms on the problem of website phishing classification. The authors compared RIPPER, C4.5 (Rules), CBA, and PRISM on a security dataset they collected containing 2500 instances and 16 features. A special hand crafted rule to collect the data was developed by the authors based on simple statistical analysis performed on the initial dataset's features. Experiments of the four rule-based classification methods showed that there are eight effective features that can be employed by the classification algorithm in combating phishing: SSL and HTTPS, Domain-age, Site-traffic, Long-URL, Request-URL Sub-domain, Multi—sub-domain, Suffix-prefix, and IP-address.

Khadi and Shinde (2013) studied the problem of email-based phishing and proposed a potential solution based on combining a RIPPER classifier with fuzzy logic. The role of fuzzy logic is to pick the main features of the email and rank them based on a probability score. Meanwhile, the role of RIPPER is to automatically use these features to classify the type of emails as ham or phishy. Two components of the email were utilized by Khadi and Shinde: the email message (spelling errors, embedded link) and URL (IP address, Length, Long URL, Suffix_Prefix, Crawler URL, Non matching URL). Moreover, very limited data consisting of just 100 instances from phishtank was in experiments involving the WEKA software tool. No comparison with other fuzzy logic or rule-based classifications was conducted by the authors. Results showed that there are twelve rules generated by RIPPER from the dataset with an 85.4% prediction rate.

Aburrous et al. (2010a) investigated rule induction methods to seek their applicability for categorising websites based on phishing features. Website features were initially manually classified into six criteria as described in an earlier report on phishing by Aburrous et al. (2008). Using

WEKA, a number of experiments with four classification algorithms (RIPPER, PART, PRISM, C4.5) were conducted against 1006 instances downloaded from Phishtank. The focus of the experiments was the classification accuracy of the classifiers produced. Results revealed that rule induction is a promising approach because it was able to detect, on average, 83% of phishing websites. The authors suggested that results obtained could be further enhanced if a careful feature selection were employed.

5.4.2.3 *Associative Classification (AC)*

The two AC methods CBA and MCAR have been evaluated on a Phishtank dataset to seek their applicability in cracking phishing (Liu et al., 1998; Thabtah et al., 2005; Abourrous et al., 2010b). Abourrous et al. (2010b) used a dataset consisting of over 1000 instances with 27 different features and applied CBA, MCAR, and four other rule-based classifiers using the WEKA DM tool. The aim was to assist security managers within organisations by building an intelligent anti-phishing tool within browsers that can detect phishing as accurately as possible. Experimental results of the six ML algorithms revealed that AC methods generated more rules than the rest of the algorithms, yet had higher predictive classifiers. More specifically, the AC systems produced showed high correlations among features linked with three major criteria: URL, Domain Identity, and Encryption. Nevertheless, the massive number of rules derived by MCAR and CBA may overwhelm end-users since they might not be able to control the anti-phishing system. Furthermore, the authors did not implement the AC rules within a browser to evaluate its real performance, which does not facilitate measuring the success or failure of their classification systems.

Recently, more domain specific AC anti-phishing systems have been created (Abdelhamid et al., 2014; Abdelhamid, 2015). These new models take into account not only two class values of the phishing problem (legitimate, phishy) but also considers a harder case to detect: the "suspicious" class label. Instances that cannot be fully classes as phishy nor as legitimate are very hard to detect by typical ML algorithms, thus increasing their false positive rates. Abdelhamid et al. (2014) and Abdelhamid (2015) have therefore enhanced current intelligent classification systems by including two distinct advantages: (1) extending the phishing problem to include suspicious cases, making it more realistic; and (2) proposing a new multi-label learning phase that can discover disjunctive in addition to conjunctive rules. These additional disjunctive rules are tossed out by existing AC methods. This new multi-label phase enhances predictive power and provides more useful knowledge to the end-user. The authors used a dataset that has 16 features and over 1500 instances, comparing the performance of their classifiers with other rule-based classifiers with respect to the knowledge derived and its accuracy. The authors employed the chi-square testing method to measure the features goodness and discriminate among features with respect to their impact on phishing. Processed data results showed high competitive performance of the new multi-label associative classifiers when compared with CBA, MCAR, rule induction, and decision trees.

5.4.2.4 *Neural Network (NN)*

One of the common ways to train a NN is trial and error (Mohammad et al., 2014a). However, this methodology has been criticised because of the time spent to tune the parameters and the requirement of an available domain expert. Thabtah et al. (2016b) proposed a NN anti-phishing model based on self-structuring the classification system rather than using trial and error. The algorithm proposed by the authors updated several parameters, like the learning rate, in a dynamic way before adding a new neuron to the hidden layer. The process of updating these NN features is performed during the building of the classification model and based on the network environment, behaviour of the desired error rate, and the computed error rate at that point. The dynamic NN model was applied to detect phishing on a large dataset from UCI containing over 11000 websites (Mohammad et al., 2015b). Experiments using different epoch sizes (100, 200, 500, 1000) have been conducted, and the results obtained exhibited better predictive systems when compared to Bayesian Network and Decision Trees.

The ANN Back Propagation algorithm (Rumelhart, et al., 1986) was investigated on a security dataset concerning website phishing by Mohammad et al. (2013). The authors collected a dataset with over 2000 instances from different legitimate and phishing sources. Processing the dataset, they tried to measure the correlation between the features and target attributes using basic univariate statistical analysis (frequency of features values and the target attribute values). Finally, they applied the Back Propagation ANN algorithm to derive anti-phishing models. The results of the study indicated that ANN is a promising approach for combatting phishing, particularly since the results showed increased accuracy of the models generated from the Back Propagation algorithm when compared with decision trees and probabilistic.

Mohammad et al. (2014a) have developed an anti-phishing NN model that relies on constantly improving the learned predictive model based on previous training experiences, Since phishers continuously update their deception methods, new features become apparent while others become insignificant. In order to cope with these changes, the authors proposed a self-structuring NN classification algorithm that deals with the vitality of phishing features. The algorithm employs validation data to track the performance of the constructed network model and make the appropriate decision based on results obtained against the validation dataset. For instance, when the achieved error against the network is lower than the minimum achieved error, the algorithm saves the network's weights and continues the training process. However, when the achieved error is larger than the minimum achieved error so far, the algorithm continues the training process without saving the weights. Other important network parameters are also updated when necessary during the building of the classification model without waiting until the model has been entirely built. Results obtained against a phishing dataset of thirty features and over 10000 instances showed that the self-structuring NN model is able to generate anti-phishing models more accurately than traditional classification approaches such as C4.5 and probabilistic approaches.

Feed Forward NN (FFNN) was applied on an email phishing classification problem by Jameel and Georg (2013). Basic implementation of a multilayer FFNN based on Back Propagation was used to differentiate suspicious from legitimate emails. Eighteen binary features were extracted from the email (header and HTML body) and made available as the training dataset attributes. These features were given values based on human rules developed by security domain experts. To derive the NN models, 6000 emails were used. The results obtained showed that FFNN is able to categorise emails with high speed and with an error rate below 2%. However, the authors have not yet embedded their FFNN into browsers for live testing.

In 2007, an experimental study contrasting five ML algorithms on the problem of classifying emails as ham or suspicious was conducted by Abu-Nimeh et al. (2007). The authors chose Classification and Regression Trees (CART), NN, Random Forests (RF), Bayesian Additive Regression Trees (BART), and Logistic Regression (LR) to measure the most successful approaches in email phishing detection. A training dataset consisting of 2889 emails and 43 email's features was used. To produce the results, the testing method employed was ten-fold cross validation and the evaluation measures used were precision, recall, and harmonic mean. Results revealed that RF achieved a lower error rate while NN generated the highest error rate among the tested classifiers. Moreover, despite RF generating the highest predictive classifiers, it derived the least false positive rate among all contrasted algorithms. The authors suggested though that more carefully chosen features may improve the performance of the anti-phishing email tool.

5.4.2.5 *Support Vector Machine (SVM)*

Proposed by Pan and Ding (2006), the SVM classification method evaluates the discrepancy between a website's identity, its HTTP transactions, and structural features. The anti-phishing solution proposed contains two layers:

- Website Identity: The set of characters appearing inside the domain name.
- Structural Features Classifier: Features that are related to the website identity and HTTP transactions.

Once a new website identity and its structural features are captured (Abnormal URL, Abnormal anchors, Server Form Handler, Abnormal certificate in SSL, Abnormal DNS, Abnormal cookies), then a SVM algorithm is trained on a historical data set consisting of the same features in order to derive the new website type. Experimental results on six features using the proposed SVM indicated that the first helps toward increasing the detection rate since malicious websites are not correlated. Furthermore, the SVM model achieved just over 83% prediction rate, and therefore more investigation is needed into the feature selection phase by including other features that could improve the performance of the classifier.

5.4.2.6 *Fuzzy Logic*

Phishing in electronic banking (Ebanking) applications has been investigated by Aburrous et al. (2008) utilizing Fuzzy Logic. A simulated phishing email was sent by the authors with the help of the security manager at Jordan Ahli Bank to measure security indicators of phishing among a sample of 120 employees after obtaining the necessary authorization (www.ahlionline.com.jo). The email urged the chosen employees to reactivate their accounts by logging in because server maintenance conducted the previous two days required account reactivations. Shocking results were obtained: 37% of the targetted employees submitted their credentials without investigation, of which 7% were Information Technology employees. The authors' goal with the simulated email was to determine features that users may look for inside the email when they suspect phishing to be used within a FL system to help in differentiating types of email.

FL has been used as an anti-phishing model to help classify websites into legitimate or phishy in Aburrous et al. (2008). The authors claimed that FL could be effective in identifying phishing activities because it provides a simple way of dealing with intervals rather than specific numeric values. Their proposed FL classification model was built manually to categorise websites using the six criteria listed in Table 5.1. Each of those criteria contains a number of phishing indicators as described in the same table. Each feature in the dataset was assigned three possible values by the authors: Phishy,

Genuine, and Doubtful. Limited results indicated that there are two effective indicators to distinguish phishiness in websites: Domain Identity and URL.

Table 5.1: Phishing Features per category (Aburrous, et al., 2008)

Criteria	N	Phishing Indicators
URL	1	IP address
	2	Abnormal request URL
	3	Abnormal URL of anchor
	4	Abnormal DNS record
	5	Abnormal URL
Encryption	1	Using SSL certificate (Padlock Icon)
	2	Certificate authority
	3	Abnormal cookie
	4	Distinguished names certificate
Source Code	1	Redirect pages
	2	Straddling attack
	3	Pharming attack
	4	OnMouseOver to hide the Link
	5	Server Form Handler (SFH)
Page Style & Contents	1	Spelling errors
	2	Copying website
	3	Using forms with *Submit* button
	4	Using pop-ups windows
	5	Disabling right-click
Web Address	1	Long URL address
	2	Replacing similar char for URL
	3	Adding a prefix or suffix
	4	Using the @ Symbol to confuse
	5	Using hexadecimal char codes

Human	1	Emphasis on security
	2	Public generic salutation
	3	Buying time to access accounts

A fuzzy based ANN model was proposed in 2015 by Nguyen et al. (2015) to classify websites based on a smaller set of phishing features related to the website's URL (PrimaryDomain, SubDomain, PathDomain) and its rank (PageRank, AlexaRank, AlexaReputation). The proposed fuzzy ANN model does not use any rules set, rather it employs a computational function to split data instances (websites) into "genuine" and "non-genuine" categories. Their model was tested against 21600 websites from legitimate and phishing sources such as Phishtank and DMOZ. They also compared the generated results with that of Aburrous et al. (2010a) and Zhang and Yuan (2008). It was discovered that their fuzzy NN model was able to slightly enhance the phishing detection rate.

5.4.2.7 *CANTINA Term Frequency Inverse Document Frequency Approach*

Carnegie Mellon Anti-phishing and Network Analysis Tool (CANTINA) is a content based anti-phishing method that determines suspicious websites using the statistical measure of Term Frequency Inverse Document Frequency (TF-IDF). Term Frequency (TF) is a statistical formula that measures keyword significance in a document while Inverse Document Frequency (IDF) measures the importance of that keyword across a large collection of documents (Witten & Frank, 2005). CANTINA evaluates the website content (links, anchor tags, forms tags, images, text, etc) for TF-IDF to produce a lexical signature of the website. This signature (top ranked TF-IDF key words) will be passed into the search engine to seek their rank in domain names and decide the type of the website. The description of the CANTINA based classification process is as follows:

1. Parse the webpage.
2. Compute the TF-IDF for the common terms of the website.
3. Select the top five terms according to the computed scores of all TF-IDF terms.
4. Add the top five terms to the URL to locate the lexical signature.
5. Input the lexical signature into a search engine.
6. Check whether the domain name of the current website matches the domain names of the top N search results (often N=30).
7. Return "Legitimate" when there is a match or "Phishy" when there is no match.

When the search results in an empty set, the current website is classified as "phishy". To overcome the "no results" problem the authors merged TF-IDF with other content features such as "IP Address," "domain age." "suspicious Images," "suspicious Link," and "suspicious URL."

Sanglerdsinlapachai and Rungsawang (2010) have used CANTINA TF-IDF, and added a few more features such as "Forms" and "Top pages' similarity linked with the domain," and removed features such as "domain age" and "known images." A dataset consisting of 200 websites was used in the experiments, and three DM methods were applied to the dataset. Results obtained, despite being limited, revealed that the reduced features set maintained a similar detection rate with that of the CANTINA features set. Moreover, adding the new features slightly enhanced the detection rate for most of the learning methods considered in the experiments.

Table 5.2 shows a brief summary of the common anti-phishing approaches that are based on automated learning along with the name of the method, the learning approached used, the first author, and their reference details.

Table 5.2 Common recent anti-phishing methods based on ML

Method name	ML technique	Reference
Dynamic rule induction	Rule induction learning	Qabajeh et al., 2014
Enhanced Dynamic rule induction	Rule induction and Covering approaches	Thabtah et al., 2016
Classification based association	AC	Aburrous, et al., 2010
Multi-label Classifier based Associative Classification	AC	Abdelhamid, et al., 2014
Self-structuring neural network	NN	Mohammad, et al., 2014
Neural Network trained with Back-Propagation	NN	Mohammad, et al., 2013
Feed Forward Neural Network	NN	Jameel & George, 2013
Fuzzy DM	Fuzzy logic	Aburrous, et al., 2008
Fuzzy DM	Fuzzy logic	Khadi & Shindi, 2014
PILFER	Decision tree	Fette, et al., 2007

Page classifier	SVM	Pan & Ding, 2006
CANTINA	Term frequency and inverse document frequency	Sanglerdsinlapachai & Rungsawang, 2010
Biased SVM, LIBSVM, ANN, Self-Organizing Map	NN, SVM and other ML techniques	Basnet, et al., 2008

5.5 Phishing Data Experimental Results

5.5.1 Experiments Settings

In deriving the predictive models, the ten-fold cross validation procedure has been adopted to compute the evaluation measures while constructing the classifiers. As discussed earlier for the UCI data experimental settings (Chapter 4) all experiments were run on a Core i5 machine with 3.1 GHz processor and 8.0 GB RAM. The minimum frequency and rule strength thresholds for eDRI were set in all experiments to 1% and 50%. The same classification algorithms used in Section 4.5 have been utilised in the phishing data experiments, including new classic non-rule based classification algorithms given the specifics of the phishing problem at hand. To be precise, we extended the algorithms to include probabilistic Bayes Net, SVM based on Sequential Minimal Optimization (SVM-SMO), and Simple Logistic (Bouckaert, 2004; Platt, 1998; Sumner et al., 2005). This is in order to evaluate the predictive power as well as the model content on the specific problem of website phishing classification. Another aim of the set of experiments is to show the advantages and limitations of the proposed eDRI algorithm when compared to other ML algorithms and rule classification algorithms used to detect phishing websites.

We have conducted two major experiments: one that does not consider feature filtering and one with feature filtering. Correlation Feature Set (CFS) filtering method was chosen due to the fact it reduces feature-to-feature correlation and maximises feature-to-class correlation (Hall, 1999). As a matter fact, CFS usually reduces the dimensionality of the input dataset significantly when compared with other known filtering and wrapping methods, such as Information Gain and Chi-Square, without drastically hindering the predictive model's performance. We also verified the filtering results by calculating the information gain scores per variable in the phishing data. The foundation of the experimental comparisons is as follows:

1) Predictive model content for decision makers, at least for rule based classifiers.

1) Predictive power measured in error rate (%).
2) Classifier size measured by the available number of rules.
3) Runtime measured by the time taken to construct the predictive models in ms.
4) Rule coverage by taking into account the number of instances covered by the rule and the available number of items in the rule's body (rule's length), as well as the earlier introduced two measures ARL and WARL (Chapter 4).
5) Number of data samples scanned to generate the classifier, to measure the increase and decrease of search space by eDRI when compared to other Covering algorithms such as PRISM.

5.5.2 Data Description and Features

The size of the security dataset includes more than 11050 examples, each one representing a website that can be either legitimate or phishy. The majority of existing features are either binary or multi value (ternary). Available website features in the dataset were extracted using a PHP script that was embedded in the web browser and then applied on Phishtank and legitimate websites between the middle of 2012 and late 2015. The data instances have been collected from the Millersmiles and Phishtank repositories (Bright 2011; PhishTank, 2011).

Before collecting the websites, a hand crafted rule was designed and then coded in PHP for each feature. For example, for the IP Address, a rule was designed that maps a website to phishy class when the IP address appears in the URL. Another rule examines URL length, and assigns a website phishy status when the URL length exceeds 75 digits. Further details on the complete description of features and their hand crafted rules can be found in Mohammad et al. (2015).

Table 5.3: Sample of twelve websites from the dataset and for eight features

IP address	URL Length	Shortening Service	@ symbol	HTTPS	Request URL	URL of Anchor	Links in Tags	Class (1- Legitimate, -1 Phishy)
-1	1	1	1	-1	1	-1	1	-1
1	1	1	1	-1	1	0	-1	-1
1	0	1	1	-1	1	0	-1	-1
1	0	1	1	-1	-1	0	0	-1
1	0	-1	1	1	1	0	0	1
-1	0	-1	1	-1	1	0	0	1
1	0	-1	1	1	-1	-1	0	-1
1	0	1	1	-1	-1	0	-1	-1
1	0	-1	1	-1	1	0	1	1
1	1	-1	1	1	1	0	1	-1
1	1	1	1	1	-1	0	0	1
1	1	-1	1	1	1	-1	-1	-1

Table 5.3 shows twelve sample websites with only eight features of the dataset for presentation purposes. The possible values (-1, 1, 0) correspond to phishy, legitimate, and suspicious values. Three features appearing in Table 5.3 are linked with ternary values, URL_Length, URL_of_Anchor, and Links in Tags while the rest are assigned binary values (1, -1). The last column in the table represents the class and is based on the twelve sample websites; four identified legitimate websites and eight phishy. Class values have been assigned based on the possible feature values per website and using the hand crafted feature rules developed in Mohammad et al. (2015) as explained above.

5.5.3 Phishing Data Results

Figure 5.4 shows the error rate generated by the rule-based predictive models and traditional classification algorithms against the phishing dataset. It is obvious that decision tree algorithms and eDRI produced the highest predictive models in regards to accuracy, and both outperformed the remaining classifiers. Particularly, eDRI achieved 0.83%, 4.79%, 4.79%, 0.69, 0.07%, 0.06% and 1.49% higher percentages of accuracy than Ridor, OneRule, ConjunctiveRule, Bayes Net, SVM-SMO, and Ada Boost algorithms, respectively. Only the C4.5 and PART algorithms outperformed eDRI on the phishing dataset, higher by 2.20% and 3.09% respectively, yet generated far more rules than eDRI (see Figure 5.6). The increase in the accuracy by C4.5 and PART was due to the fact that these algorithms utilise information gain metrics to choose the root variable in a repetitive manner. Since we have several variables with ternary values in the input dataset, when these variables are evaluated by decision tree algorithms a branch is made and each path from the root node to any leaf denotes a rule. Therefore, models generated by C4.5 and PART on the phishing dataset are very large with respect to the number of rules derived when compared with the remaining classifiers, which may limit their use in real domain applications including the dealt with here: phishing.

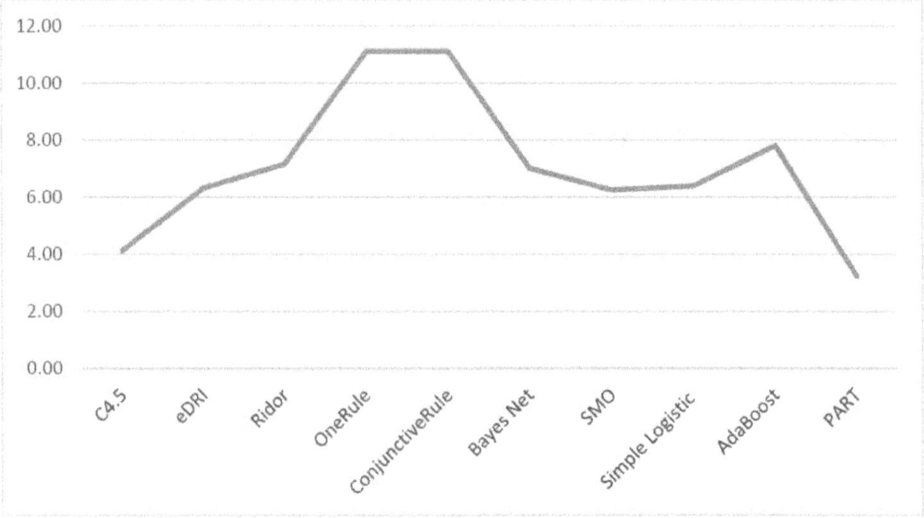

Figure 5.4 The One error rate (%) of the considered algorithms on the phishing dataset

The fact that eDRI was able to achieve consistently lower error rates than powerful classic algorithms such as Naïve Bayes, SVM-SMO, and Simple Logistic as well as rule-based classfiers such as OneRule, Ridor, Conjunctive Rule, and PRISM, is clear evidence that the eDRI algorithm learning strategy has improved the classifier's predictive power. In fact, the ability to limit the use of the default class by eDRI has increased the overall accuracy of the classifier. Despite the increased accuracy rate by the C4.5 and PART algorithms, they discovered more rules than their counterparts. To be precise, the PART algorithm extracted 130 and 144 additional rules more than eDRI and RIDOR, respectively. C4.5 was worse, having generated from the phishing dataset 268 and 258 more rules than eDRI and RIDOR algorithms, respectively. These massive classifiers may slightly increase classification accuracy yet are of no practical use for decision makers since they are hard to understand, maintain, and manage. In other words, the additional knowledge may contribute to increase C4.5 and PART predictive performance, but results in uncontrollable classifiers that may overwhelm end-users. Overall, classification algorithms that derived rule-based classifiers other than C4.5 and PART, more specifically eDRI and RIDOR, were able to produce predictive models with good levels of accuracy and maintainable sets of rules.

Phishing fetaures in the input dataset were assessed to determine the smallest set of effective phishing features. Table 5.4 displays the top features after applying the CFS filtering method, and the results were verified based on information gain scores. Nine vital features for phishing detection were chosen using the CFS and information gain filtering methods. The best scored feature was SSL_Final_State followed by URL_Of_Anchor. As a result, it was decided to run the classification algorithms and derived predictive models from the nine feature sets resulting after application of the CFS filtering method. All remaining features are insignificant to the class variable in the dataset, thus having a limited effect on phishing detection, and therefore they have been ignored.

Table 5.4: Phishing features derived by Correlation Features Set filtering method along with their information gain scores

Phishing Feature Name	Information Gain Score
Prefix_Suffix	0.1230
having_Sub_Domain	0.1090
SSLfinal_State	0.4994
Request_URL	0.0460
URL_of_Anchor	0.4770
Links_in_tags	0.0470
SFH	0.0370
web_traffic	0.1145
Google_Index	0.0110

Figure 5.5 shows the error rate (%) results of all considered algorithms against the top nine features listed in Table 5.4. These figures indicate strong and consistent performance in regard to predictive power by eDRI, PART and C4.5 algorithms. Specifically, the eDRI algorithm achieved the least error rate on the reduced phishing data among the rule induction algorithms, while C4.5 and PART algorithms derived the highest quantity of

predictive classifiers. Moreover, pruning useless and redundant rules have an obvious positive effect on performance by ensuring only effective rules are utilised in the prediction. These rules, unlike other algorithms, cover a larger portion of the training dataset. Decision tree algorithms (C4.5, PART) produced the least error rate, yet still produced very large predictive models that are associated with an excessive number of rules. To be precise, and for the nine feature dataset, C4.5 and PART derived 60 and 59 more rules, respectively, than eDRI.

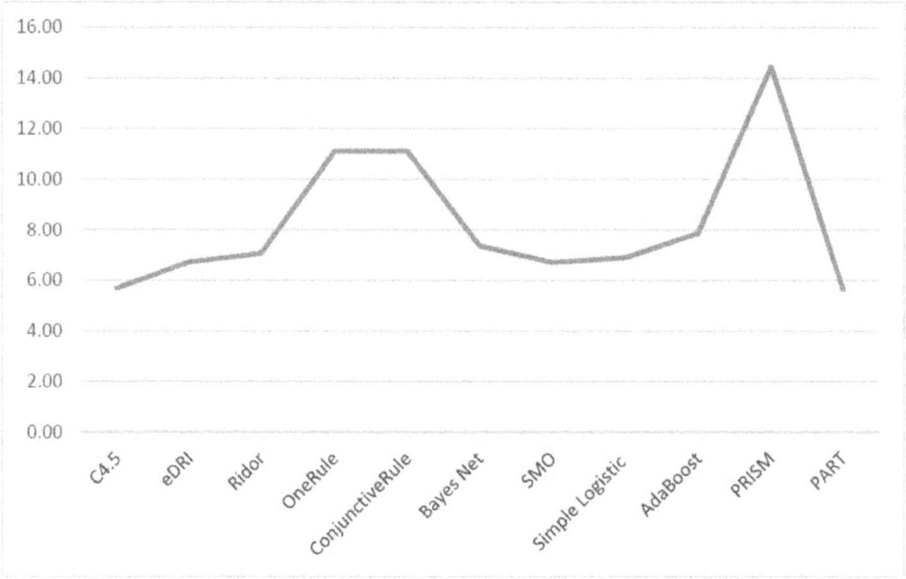

Figure 5.5. The One error rate (%) of the considered algorithms on the reduced nine features phishing set chosen by CFS filtering

We looked at the number of rules eDRI and the remaining rule-based classfiers derive from the complete phishing dataset and the nine-feature dataset, respectively. The results, shown in Figures 5.6 and 5.7, illustrate that the proposed eDRI algorithm substantially reduced the classifier size in comparison to the considered rule-based algorithms. In fact, eDRI generated models with a lower number of rules than the remaining algorithms from both the complete and reduced phishing dataset. For instance, classic Covering algorithms such as PRISM derived 325 more rules than eDRI without achieving any performance gain in classification accuracy. Many of PRISM's rules cover very limited data samples, which makes them overfit the training dataset.

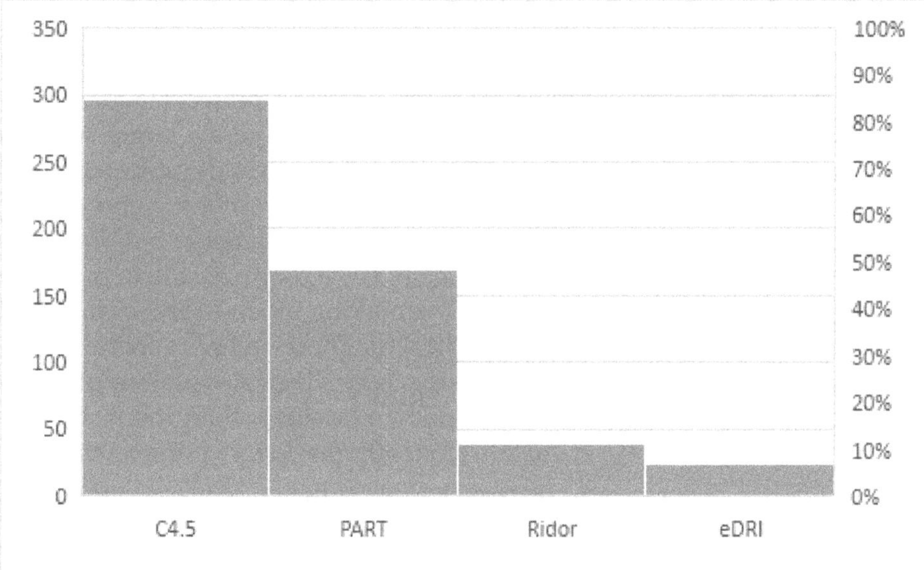

Figure 5.6. Number of rules generated by the rule-based algorithms from the phishing data

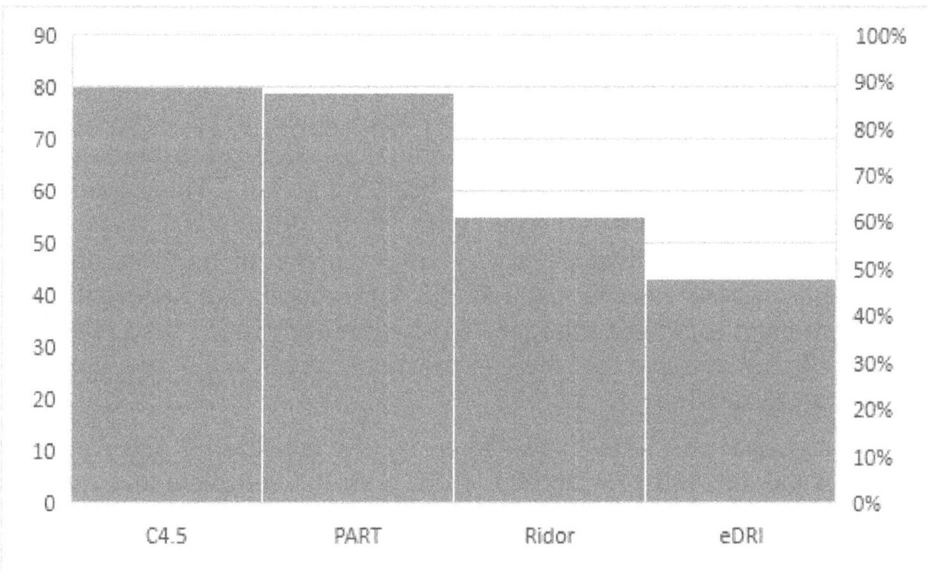

Figure 5.7. Number of rules generated by the rule-based algorithms from the nine features phishing data

We looked further into the classifier produced by eDRI from the phishing dataset and observed that some of its rules have errors greater than zero. Despite the fact that these rules do not hold an accuracy of 100%, they are useful knowledge that the PRISM classifier is unable to provide because it only poduces perferct rules with low data coverage. This indicates that generating rules which are not necessarily perfect (100% accuracy) not only minimises the classifier size but also covers a larger number of training data examples per rule. Therefore, the overall performance of the classifier gets enhanced in regard to predictive accuracy and human controllability. In other words, having a smaller classifier is a definite advantage for managers and decision makers since they are able to process a lower amount of data during the decision making process. These classifiers work fine for applications such as medical diagnoses where general practitioners can enjoy a concise set of rules for daily diagnoses of their patients. Overall, eDRI consistently derived smaller classifiers than the rest of the classification algorithms, yet sustained steady predictive performance.

The relationship between the rule length in the classifier and the number of instances the rule was covered by the Covering algorithms (PRISM, eDRI) was also investigated. The proposed average rule length (ARL) and weighted average rule length (WARL) (Chapter 4) were used. For the dataset examined, PRISM's classifier had associated ARL and WARL values of 4.25 and 7.29 respectively; meanwhile eDRI's classifier ARL and WARL were 4.95 and 5.16, respectively. WARL measures how the training attributes have been utilised to build the classifier. As such, it is clear that PRISM needed more features (two more features on average) than eDRI to make the rules, without achieving any enhancement in predictive performance. As a matter of fact, PRISM excessively searched for more items to be added per rule during the classifier building process. This explains the many specific rules (rules associated with a large number of items) derived by PRISM, which classifies each very small portion of the training dataset though in many cases is just a single data sample. We believe that the above rationale is behind a larger WARLvalue for PRISM than eDRI. On the other hand, it seems that there is consistency between both algorithms with respect to ARL.

We analysed the time in ms taken by the considered algorithms to construct the anti-phishing models from both the complete and reduced datasets respectively, which is shown in Figures 5.8 and 5.9. Simple Logistic and SMO-SVM were excluded since both took longer than expected to build the models with respect to time in ms: Simple Logistic spent 3691 and 708 ms, while SMO-SVM spent 4327 and 1383 ms in building the models, respectively. Thus, both of these predictive algorithms proved very slow in mining the phishing data.

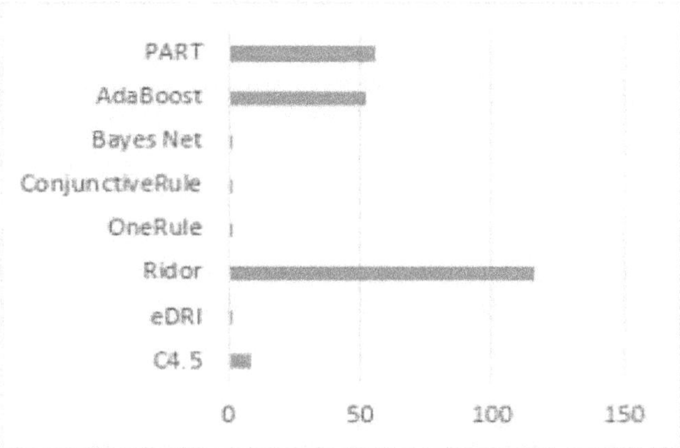

Figure 5.8 Time in ms taken to build the models by the considered algorithms from the security data

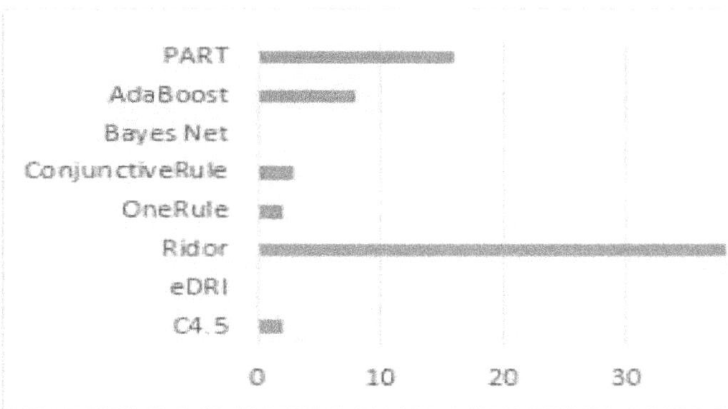

Figure 5.9 Time in ms taken to build the models by the considered algorithms from the reduced nine variables security data

Data processing time figures clearly indicated that eDRI is the most efficient algorithm among all those considered for detecting phishing activities, followed by Bayes Net. In addition, the OneRule and ConjunctiveRule algorithms were able to generate models quicker than decision trees due to the simplicity in their rule induction mechanisms. Overall, Ridor was the slowest rule induction algorithm followed by decision tree algorithms (PART, C4.5). The proposed eDRI algorithm showed consistently more efficient classifiers than the remaining algorithms, for both the complete and the reduced nine features dataset. This can be attributed to the efficient rule induction strategy employed which automatically generates rules and dynamically updates the remaining potential rule rank. Resultant is the integration of the classifier building phase with the training phase in eDRI, thus saving computing resources.

We also investigated the number of times eDRI passed over training data examples when compared to PRISM. The number of rows that both algorithms have to scan while building rules during the learning process were recorded. It turns out that PRISM passed over 34,465,178 training examples during the process of creating only 397 rules. Meanwhile eDRI scanned 9,388,208 data examples resulting in 25 rules. These results show that the proposed algorithm has substantially reduced the search space for rules. The fact that PRISM has to pass over many more millions of data examples shows the poor mechanism employed by this algorithm which over fits the training data while aiming to induce perfect yet low data coverage rules. This mechanism requires overtraining by repetitively splitting data examples whenever an item has to be added to a rule in order to increase the rule's accuracy. The outcome guarantees the production of rules having zero error, but it is time consuming and demanding of computing resources. The frequency threshold employed in the eDRI algorithm was able to successfully discard many infrequent items that are associated with low data representation. This minimised the search space of items, therefore the classifier size was reduced. Moreover, allowing rules which are not 100% accurate to be generated decreased the number of data examples that eDRI had to pass over. The proposed eDRI algorithm does not always generate rules that are perfect but rather generates rules with large data coverage and only acceptable accuracy as it was obvious from its classifier containing a much lower number of rules.

5.6 Chapter Summary

In this chapter, the history of the phishing problem has been described along with the main phases in its lifecycle. Anti-phishing methods that depend on legal, educational, and user experiences have been reviewed along with their associated advantages and disadvantages. The main focus of the Chapter, however, has been on ML anti-phishing techniques based on decision trees, SVM, NN, rule induction, AC, and computational intelligence. Phishing was then investigated as a classification problem. The experimental study on real phishing datasets was conducted in Section 5 to explore the applicability of the proposed eDRI algorithm as an anti-phishing tool and how it performs against other DM algorithms. Experiments were performed in two situations:

a) Complete thirty feature real phishing datasets

b) Applying the CFS filtering method to identify effective features and measure the impact on the predictive model's accuracy

The aim of these experiments was to evaluate how effective rule based models, and in particular eDRI, are in classifying websites and producing useful knowledge for decision makers. Classification accuracy, runtime and the number of rules were the measures of evaluation of eDRI with respect to the following algorithms: decision trees C4.5, OneRule, Conjunctive Rule, Bayes Net, Support Vector Machine (SMO), Simple Logistic, Boosting (AdaBoost), PART, Ridor, and eDRI. The experimental results showed that

1) eDRI is an efficient rule-based method that normally generates anti-phishing models quicker than other rule-based algorithms and classic DM methods.
2) eDRI derives high quality models with respect to error rate, and outperformed all rule-based classifiers. Moreover, decision tree algorithms derives higher predictive models than eDRI yet with excessive additional rules, make them difficult to utilise as a decision making tool and thus limiting their use.
3) The number of rules in eDRI models are consistently smaller than most of the considered algorithms, therefore eDRI models are more appropriate knowledge base for phishing classification.
4) Nine features with high information gain scores were selected by CFS as the highest influential phishing features. The SSL_Final_State and URL_Of_Anchor features were specifically the ones with more presence in the rule-based classifiers as well as with higher information gain scores.
5) The highest ranked rules in eDRI, PRISM, RIDOR, C4.5, and PART are associated with SSL_Final_State and URL_Of_Anchor features.
6) eDRI showed superiority over the remaining algorithms, other than decision trees with respect to the predictive accuracy of models derived from the nine features dataset.

The PhD dissertation conclusions and future research directions are discussed in the following chapter.

Chapter 7

Conclusions and Future Work

6.1 Research Summary

Key research issues related to the utilization of the Covering approach in classification have been investigated. A number of improvements to the training phase, efficiency, and classification steps in the Covering approach have been presented. The final outcome is new dynamic rule induction algorithm, the eDRI algorithm, with novel training, rule pruning, and class prediction procedures. Experimental results have shown the ability of eDRI's Covering methodology to produce efficient and predictive accuracy in classification. Following a thorough evaluation of the eDRI algorithm on generic classification datasets and domain specific data related to web security, the different improvements presented in this PhD dissertation have been peer reviewed and presented at IEEE conferences, and published in reputable data analysis journals such as *Expert system with Application* and *Management Analytics*.

6.2 Research Contributions

The following main research contributions are summarised in this section.

1. Dynamic Rule Discovery

One of the major hurdles that classic Covering algorithms face is the need to repeatedly scan part of the training dataset whenever a rule is constructed. Indeed, for the first rule to be derived a complete scan over the training dataset is usually requires; once the rule is derived and its training data instances discarded, a complete scan of the remaining data instances is required to produce the second rule and so forth. These repetitive data scans are computationally very expensive and demand high computing resources, particularly runtime, to create a predictive model. To deal with this issue, the proposed eDRI algorithm dynamically updates frequencies of the attribute values and potential rules whenever a rule is produced. In particular, eDRI employs a data structure that minimises passing over unnecessary data instances by storing attribute values in a special data structure. This data structure is actively updated whenever a rule is generated by decrementing the frequencies of all impacted attribute values without the need to keep reading training data instances.

A number of experiments, using twenty datasets belonging to various application domains, showed that eDRI is faster than both Covering and rule induction algorithms when building predictive models. Specifically, eDRI was consistently more efficient than Ridor, RIPPER, PRISM, and C4.5 algorithms on each of the twenty datasets. This can be attributed as mentioned above to the dynamic training procedure that eDRI adopts, which limits the scanning of training data instances and relies on a special data structure that holds the necessary information to compute the rules' strengths and attribute value frequencies. Additional evidence supporting the efficiency of the proposed eDRI algorithm was obtained via a thorough analysis on phishing application data, which showed that that eDRI was the fastest algorithm among the rule-based and classic machine learning classifiers. In addition, eDRI was compared to the classic Covering algorithm PRISM; the results showed that eDRI scans 5 times less data examples than PRISM to produce their corresponding predictive models.

2. Enhancing the Performance of the Classification Models (Classifiers)

Three specific improvements related to Covering classification performance have been achieved:

A. Cutting down the items' search space during the training phase

The proposed eDRI algorithm adopts a frequency threshold parameter that substantially reduces the search space for attribute values during the training phase. This parameter distinguishes attribute values that have a chance to make rules (strong attribute values) from others that have a low statistical significance (weak attribute values) in the input dataset during the training phase. This frequency threshold parameter is used to eliminate weak attribute values as early as possible by testing the actual frequency of attribute values whenever a rule is generated. This was corroborated experimentally in Chapter 4, which also was reported in Qabajeh et al. (2015), with an improvement on computing resources, especially run time reduction and the number of passes required to build the classifier.

B. Minimising the number of rules for better controllability

A pruning procedure that removes rules having an insufficient strength in a dynamic manner has been proposed. The eDRI rule pruning procedure discards rules whenever they are unable to pass the required rule_strength parameter. Moreover, when a rule is produced, and the remaining attribute value frequencies are decremented because of the rule's data removal, eDRI prunes other potential rules in real time if their updated strengths fall below the required minimum strength. Thus, rules that are not only statistically representative in the remaining training data but also have a higher rank in the classifier are kept. This rule pruning procedure significantly decreases the number of rules in the classifiers without hindering the predictive accuracy, as experiments on twenty datasets reported in Chapter 4 and 5 revealed. Indeed, the number of rules derived by the proposed eDRI algorithm is normally smaller than other Covering and decision trees algorithms, such as PRISM and C4.5. More importantly, the classifiers generated from the phishing dataset reported in Chapter 5 showed a fewer number of rules for eDRI than for C4.5, PART, PRISM, and Ridor. Specifically, PART and C4.5 algorithms generated 130 and 268 more rules than eDRI, respectively, which has the obvious consequence of making their models difficult to manage and understand by end-users. Even the rule induction algorithm Ridor generated more rules than eDRI on the phishing dataset.

C. Avoiding Over fitting and Under fitting, and Improving Class Prediction

Rule learning in Covering classification should be restricted so that the discovered models will not over fit the training dataset. For example, in rule induction, RIPPER stops learning when the rule achieves an error rate above 50%. Classic Covering and rule induction approaches like PRISM and REP over fit the training dataset because they continue adding items to the rule until it becomes perfect (0% error rate). The proposed eDRI algorithm minimises over fitting by allowing a small window of errors to some rules as long as:

1) They have a strength above the minimum required rule strength,
2) The attribute values in their bodies have acceptable training data representation, and
3) They cover large training data examples.

The above three conditions together act as a quality assurance strategy to maintain rule generation while reducing over fitting. One of the possible limitations to occur with the eDRI algorithm is associated with setting the rule strength to 100% when the process of making a rule can become excessive. This problem has been minimised by allowing the user to determine the rule strength according to the sensitivity of the domain data. Warming up experiments using a large number of values for the rule strength and frequency thresholds have showed that the frequency threshold can be set to 1-2% in order to balance the classifier size and predictive accuracy (Thabtah et al., 2015; Abdelhamid, 2015). In addition, the results in (Thabtah et al., 2015; Abdelhamid, 2015) showed that the rule_strength threshold does not directly influence model performance, so it can be set to any value based on users' experience and the application data characteristics.

To improve the classification step, we examined two separate approaches; one rule and multiple rules prediction (Thabtah, et al., 2011). We found that the differences between both approaches are insignificant, yet when the multiple rules approach is adopted for the classification of test data, this may reduce usage of the default class rule, and therefore slightly enhance the predictive accuracy of the models. For this reason, we developed a new prediction method (Chapter 3) that improved classification accuracy for some of the datasets considered in chapters 4 and 5.

3. Rules Set as an Effective Anti-Phishing Tool: A Case Study

Phishing is a web threat that is continuously evolving. Phishing creates authentic websites are supplanted by purposely created malicious websites to deceive online users.

A predictive ML approach to phishing has been extensively presented in Chapter five, in addition to a critical analysis of the advantages and disadvantages of traditional educational, training, legal, and awareness solutions to it. Recent intelligent anti-phishing approaches based on computational intelligence and ML, such as SVM, NN, Fuzzy Logic, AC, CANTINA, etc., have been surveyed, with both their the impact on detecting phishy websites and their advantages and drawbacks being identified.

Chapter five presented as well an important comprehensive experimental study on a real phishing dataset that consisted of over 11000 websites from different sources such as Millersmiles and Phishtank. A wide range of ML algorithms and the proposed eDRI algorithm were applied to this dataset which aimed at identifying whether eDRI is an effective anti-phishing detector. To such aim, a wide range of different performance metrics were used, including phishing detection rate, runtime, classifier size and knowledge presented to the decision maker. The experiment results showed that eDRI is indeed an effective and efficient anti-phishing algorithm for detecting phishing websites. The proposed eDRI algorithm achieved higher accuracy than rule-based induction, probabilistic, boosting, SVM and even AC ML algorithms. Specifically, eDRI achieved 0.83%, 4.79%, 4.79%, 0.69, 0.07%, 0.06% and 1.49% higher accuracy than Ridor, OneRule, ConjunctiveRule, Bayes Net, SVM-SMO, and Ada Boost algorithms, respectively. Furthermore, the experimental results also showed that smaller sets of effective rules were produced by eDRI that can explain in detail the associations among website features that have increased influence on detecting phishing activities. Finally, when using filtering methods such as CFS, eDRI maintained its competitive performance even when the dimensionality of the phishing data was significantly reduced to nine features.

6.3 Future Work

The following challenges are considered for future avenues to research.

1. Data Representation for Fast Training Phase

There are two main approaches for presenting data: horizontal and vertical format. Data in the horizontal format used in this PhD dissertation consists of a number of rows and columns and similar to the relational database approach. In each data row, a number of attribute values are provided. In reading a horizontal training dataset, the learning algorithm usually moves sequentially, counting the number of occurrences for each attribute value and the class value. These numbers are used to weigh the rule significance and are often used within mathematical or statistical rule stopping formulas such as information gain, foil gain, expected accuracy, or rule strength. Unfortunately, the learning algorithm must perform multiple passes over part of or the complete training dataset, especially in Covering classification. These repetitive data scans negatively affect the efficiency of the data processing phase of the learning algorithm.

The other data format, rarely used within Covering algorithms, is the vertical data format. Using this format, simple yet effective intersections among item locations in the training dataset can be performed. This approach is popular in association rule mining, and several algorithms such as DCLAT, CHARM, and MAC have adopted it for fast data processing (Zaki & Gouda, 2003; Zaki & Hsiao, 2002; Abdelhamid, et al., 2012). The vertical data format is believed to have the potential to enhance the efficiency of the exhaustive search for rules in covering and other rule-based classifications such as rule induction. This format could enable algorithms like eDRI, and others such as RIPPER and RIDOR, to limit scanning of the database attribute values. Experimental research is needed to ascertain or corroborate this research hypothesis.

2. Partly Matching Class Forecasting Methods

In classic Covering classification, the process of classifying a test instance is primarily based on using the first rule that fully matches the test instances' attribute values. Often, the learning algorithm evaluates whether a rule's body is contained inside the test instance, and if so, allocates the class label of that rule to the test instance. However, there are scenarios when no rule is fully contained in the test instance, and in these cases the learning algorithm selects the default rule to assign the class to the test instance. Unfortunately, this approach may increase the number of misclassifications, and therefore might hinder the predictive accuracy of the model because default class rule is a weak rule that is derived from the remaining unclassified training data during the rule discovery phase. Therefore, the choice of the default rule and the way it is constructed seems to be arbitrary rather than based on an intelligent based approach.

The improvement of the class-forecasting phase in Covering classification seems to essentially require for partly matching of rules to play a role in classifying test instances rather than applying the default class rule. For example, assuming that a test instance needed a class with no rules fully contained in it, the learning algorithm could then seek for all partly matching rules, i.e. rules with at least one-attribute values within the test instance. Once those rules are determined, the learning algorithm could cluster them into groups and a voting mechanism could be used to allocate the class, such as the class that belongs to the cluster with the largest number of rules or that has the largest average rules strength value (Thabtah et al. 2011). This strategy of class forecasting may decrease the number of classifications since it limits the number of times the default class is utilised in class predictions. Thus, the overall predictive accuracy of the model could be improved. Experimental research is needed to ascertain or corroborate this research hypothesis.

3. Distributed Covering

As a result of the literature review and critical analysis conducted in Chapter two, we can conclude that most Covering and rule induction algorithms are time consuming because of their exhaustive data processing and rule pruning phases. For instance, REP generates a large rule set that is improved it by pruning each rule. To do that, each item removal within the rule is examined to determine whether rule accuracy can be enhanced for all rules that have been generated. RIPPER and IREP eliminate the generation of a complete set of rules, deriving them instead one at a time. Both algorithms adopt extensive pruning based on growing and pruning datasets. Moreover, similarly to REP, each rule must also be taken for evaluation once it has been generated. These learning approaches have been criticized for being slow when processing data locally. The problem becomes more severe in real time applications that have data distributed over different locations. Therefore, new approaches of mining distributed data and big data are needed.

Spark and MapReduce seem appropriate parallel and distributed platforms for data processing in these situations (Hammoud, 2010; Taylor, 2008). These platforms are becoming increasingly popular for mining large scale training datasets that are scattered over many locations. This is due to their efficiency, simplicity, and effectiveness for data processing in a parallel manner. For instance, MapReduce has been adopted by many organizations such as Google, Yahoo, and Amazon to construct petabyte data centers. These centers consist of hundreds of thousands of nodes with hardware and software capabilities that allow efficient parallel data analysis. A need exists to amend current Covering algorithms to handle big data, especially in domains that generate datasets with large dimensionality like Bioinformatics, text mining, and medicine.

6.4 Limitation of the Proposal and Potential Improvements

This section explains the limitation of the proposed Covering algorithm and potential ways to improve it.

Limitations and Potential Improvements

1. eDRI algorithm works only locally so if the datasets are scattered all over different locations this can be a challenge. In today's era of cloud computing and affordable computing resources such as processing clusters, bandwidth and distributed platforms eDRI can be extended to handle datasets that are not only very big but also distributed on different locations. We recommended SPARK or Hadoop platforms for a distributed version of eDRI to handle scattered datasets for parallel processing.
2. Unstructured, and multimedia applications' data are increasing in the last few years. Companies are collecting different types of variables in forms of audio, video and images. In the current version of eDRI, these types of datasets are not considered because of the complex nature, which requires extensive and expensive pre-processing.

   There are good potential that eDRI learning mechanism to be linked with deep learning research to handle one or more of the challenging datasets.
3. One possible improvement on eDRI is adopting forward pruning procedures such as pessimistic error estimation into its learning phase in order to reduce further weak rules. Moreover, we can also

integrate backward pruning procedures such as sub tree replacement to increase the chance of incorporating only variables that have low correlation with each others but high correlation with the class labels. The results are expected to be more concise classification systems with less number of variables.

4. The experimental methodology discussed in Chapter 4 was based on utilizing the standard evaluation measures used by other scholars in predictive tasks within data mining and machine learning. In particular, we have used tenfold cross validation method as a testing methodology. This method ensures that the sample data examples used during the training phase is different than the data examples used during the classifiers' evaluation phase. This consequently sustains fairness and legitimacy in the classifiers' performance of the different data mining algorithms utilized during the experimentations.

   Alternative non-parametric statistical analysis approaches can be potentially useful in measuring the significance of the results achieved by a data mining method in comparison with other methods. The application of these techniques is possible, yet these alternative techniques might produce a completely different type of data analysis to the one achieved. Non-parametric testing approaches are normally named distribution-free tests since they have no pre-assumption that the data used follows a certain distribution in statistics (Gravetter & Wallnau, 2016; Zhang & Zhou, 2005). These tests can be used by comparing between learning algorithms in parwise or in groups (SCI2S, 2017). For instance, Wilcoxon signed-rank-test (SCI2S, 2017) can be adopted to figure out the performance of two data mining algorithms by individual comparing the accuracy performance of the selected algorithms.

   Overall, there are several useful non-parametric tests that can be employed to verify the performance results of the classifiers including but not limited to

   - The Iman and Davenport Test
   - Wilcoxon test
   - Friedman test
   - Quade Test

A investigation of non-parametric test can be potentially useful to reveal the statistical significance of the classifiers' performance results. Nevertheless, the sample dataset size should be large enough in order to compare between paired to multiple classifiers since the sample size ensures the power of the statistical test is sustainable. More information about common non-parametric tests and other statistical procedures such as hypothesis testing can be found at (SCI2S, 2017; Gravetter & Wallnau, 2016).

Bibliography

1. Aaron, G., Manning, R. (2014) APWG Phishing Reports. http://docs.apwg.org/reports/apwg_trends_report_q4_2014.pdf [Accessed March 20 2016].

2. Abdehamid N. (2015) Multi-label rules for phishing classification. Applied Computing and Informatics 11 (1), 29-46.

3. Abdelhamid, N., Jabbar, A. A., & Thabtah, F. (2016, August). Associative Classification Common Research Challenges. In Parallel Processing Workshops (ICPPW), 2016 45th International Conference on (pp. 432-437). IEEE.

4. Abdelhamid N., Thabtah F., Ayesh A. (2014) Phishing detection based associative classification data mining. Expert systems with Applications Journal. 41 (2014) 5948–5959.

5. Abdelhamid N., Thabtah F., (2014) Associative Classification Approaches: Review and Comparison. Journal of Information and Knowledge Management (JIKM). Vol. 13, No. 3 (2014) 1450027.

6. Abdelhamid N., Ayesh A., Thabtah F., Ahmadi S., Hadi W (2012b) MAC: A multiclass associative classification algorithm. Journal of Information and Knowledge Management (JIKM). 11 (2), 1250011-1 - 1250011-10. Worldscinet.

7. Abu-Nimeh, S., Nappa, D., Wang, X. and Nair (2007) A Comparison of Machine Learning Techniques for Phishing Detection. In The 2nd annual Anti-Phishing Working Groupse Crime researchers, eCrime '07. New York, NY, USA, 2007. ACM.

8. Aburrous M.., Hossain M., Dahal K.P. and Thabtah F. (2010A) Experimental Case Studies for Investigating E-Banking Phishing Techniques and Attack Strategies. Journal of Cognitive Computation, Springer Verlag, 2 (3): 242-253.

9. Aburrous M., Hossain M., Dahal K.P. and Thabtah F. (2010B) Asso-

ciative Classification techniques for predicting e-banking phishing websites. Proceedings of the 2010 International Conference on Information Technology, Las Vegas, Nevada, USA, 2010, pp. 176-181.

10. Aburrous M., Hossain A., Dahal K., Thabtah F. (2008) Intelligent Quality Performance Assessment for E-Banking Security using Fuzzy Logic. Proceedings of the 7th IEEE International Conference on Information Technology (ITNG 2008). Las Vegas, USA.

11. Afroz, & Greenstadt, R. (2011) PhishZoo: Detecting Phishing Websites by Looking at Them. In Fifth International Conference on Semantic Computing (September 18-September 21). Palo Alto, California USA, 2011. IEEE.

12. Aha, D.W., Kibler, D. and Albert, M.K. (1991). Instance-based learning algorithms. Machine Learning, 6,pages 37-66.

13. Ng, A. Y.; Jordan, M. I. (2002). On discriminative vs. generative classifiers: A comparison of logistic regression and naive Bayes. NIPS.

14. Arachchilage NAG, S Love S. (2013) A game design framework for avoiding phishing attacks. Computers in Human Behavior 29 (3), 706-714

15. Arachchilage NAG., Cole M. (2011) Design a mobile game for home computer users to prevent from "phishing attacks". Information Society (i-Society), 2011 International Conference on, 485-489.

16. Arachchilage NAG,, Rhee Y., Sheng S., Hasan SH., Acquisti A., Cranor L. F., Hong J. (2007) Getting users to pay attention to anti-phishing education: evaluation of retention and transfer. In eCrime '07 Proceedings of the anti-phishing working groups 2nd annual eCrime researchers summit. Pittsburgh, PA, USA, 2007. ACM.

17. Ayyat Susan, Lu J., Thabtah F. (2014) Class Strength Prediction Method for Associative Classification. Proceedings of the IMMM 2014, The Fourth International Conference on Advances in Information Mining and Management, pp. 5-10. Paris, France, July 2014. Best Paper Award.

18. BBC News, 2005.http://news.bbc.co.uk/2/hi/uk_news/england/lancashire/4396914.stm [Accessed 11 April 2016].

19. Basnet R., Mukkamala S., Sung AH (2008) Detection of phishing attacks: A machine learning approach (2008) Soft Computing Applications Industry, pp. 373-383.

20. Bouckaert, R. R., Frank, E., Hall, M., Kirkby, R., Reutemann, P., Seewald, A., and Scuse, D. (2008). WEKA Manual for version 3.6.0. University of Waikato, Hamilton, New Zealand, 2008.

21. Bouckaert, R. R., Frank, E., Hall, M., Kirkby, R., Reutemann, P., Seewald, A., and Scuse, D. (2013). WEKA Manual for version 3.7.8. University of Waikato, Hamilton, New Zealand, 2013.

22. Bouckaert, R. R., (2004). Bayesian network classifiers in Weka. (Working paper series. University of Waikato, Department of Computer Science. No. 14/2004). Hamilton, New Zealand: University of Waikato.

23. Boser, B. E.; Guyon, I. M.; Vapnik, V. N. (1992). A training algorithm for optimal margin classifiers. Proceedings of the fifth annual workshop on Computational learning theory – COLT '92. p. 144. doi:10.1145/130385.130401. ISBN 089791497X.

24. Bramer, M.A. (2000) Automatic induction of classification rules from examples using N-PRISM", Research and Development in Intelligent Systems XVI, Springer-Verlag, pp. 99–121, 2000.

25. Bramer A. (2002) An information-theoretic approach to the pre-pruning of classification rules, in: B. N. M Musen, R. Studer (Eds.), Intelligent Information Processing, Kluwer, 2002, pp. 201{212}.

26. Breiman L., Friedman J. H., Olshen R. A., and Stone C. J. (1984). Classification and Regression Trees. Chapman and Hall/CRC, Boca Raton, FL, 1984.

27. Bright, M. (2011) MillerSmiles. [Online] Available at: http://www.millersmiles.co.uk/ [Accessed 09 January 2016].

28. Brookhouse, J., & Otero, F. E. (2016, July). Using an Ant Colony Optimization Algorithm for Monotonic Regression Rule Discovery. In Proceedings of the 2016 on Genetic and Evolutionary Computation Conference (pp. 437-444). ACM.

29. Caruana, R.; Niculescu-Mizil, A. (2006). An empirical comparison of supervised learning algorithms. Proc. 23rd International Conference on Machine Learning. CiteSeerX 10.1.1.122.5901Freely accessible.

30. Cendrowska, J. (1987) PRISM: An algorithm for inducing modular rules. International Journal of Man-Machine Studies, Vol.27, No.4, 349-370.

31. Chen, J. & Guo, C. (2006) Online Detection and Prevention of Phishing Attacks (Invited Paper). In First International Conference on Communications and Networking in China. ChinaCom '06. Beijing, 2006. IEEE.

32. Clark P. and Niblett T. (1989). The CN2 induction algorithm. Machine Learning, 3:261–283, 1989.

33. Clark P. and Boswell R. (1991). Rule induction with CN2: Some recent improvements. In Proceedings of the 5th European Working Session on Learning (EWSL-91), pp. 151–163, 1991.

34. ClickDimensions (2014) www.clickdimensions.com/sites/default/files/PDF/WhitePaper-CASL.pdf [Accessed 12 May 2016].

35. Cloudmark Org. (2002) Cloudmark. http://www.cloudmark.com/en/home [Accessed 10 February 2016].

36. Cohen, W., 1995. Fast Effective Rule Induction. In In Proceedings of the Twelfth International Conference on Machine Learning. Tahoe City, California, 1995. Morgan Kaufmann.

37. Cohen W. and Singer Y. (1999). A simple, fast and effective rule learner. In Proc. 16th Nat. Conf. on Artificial Intelligence, pages 335–342. AAAI/MITPress, 1999.

38. Coulter M. (2013) Strategic Management in Action. 6th Edition, Pearson Education.

39. Cortes, C., and Vapnik, V. (1995). Support-Vector Networks. *Machine Learning, 20* (3), 273 – 297.

40. Cost, S. and Salzberg, S. (1993). A weighted nearest neighbor algorithm for learning with symbolic features.Machine Learning, 10, pages 57-78.

41. Cover, T.M. and Hart, P.E. (1967). Nearest neighbor pattern classification. IEEE Transactions on Information Theory, 13, pages 21-27.

42. Dasha R., Dashb P. K. (2016). A hybrid stock trading framework integrating technical analysis with machine learning technique. The Journal of Finance and Data ScienceVolume 2, Issue 1, March 2016, Pages 42–57.

43. Dain, O., Cunningham, R., and Boyer, S. (2004). Irep++, a faster rule learning algorithm. Proceedings of the Fourth SIAM International Conference on Data Mining.

44. Demšar, J. (2006). Statistical comparisons of classifiers over multiple data sets. Journal of Machine learning research, 7(Jan), 1-30.

45. Diebold, F.X. and Mariano, R. (1995), Comparing Predictive Accuracy. Journal of Business and Economic Statistics, 13, 253-265.

46. Dhamija, R. and Tygar, J. (2005) The battle against phishing: Dynamic Security Skins. In The 1st Symposium On Usable Privacy and Security. New York, NY, USA, 2005. ACM Press.

47. Domingos, P. (1996a). Unifying Instance-Based and Rule-Based Induction. Machine Learning 24, 141–168 (1996)

48. Domingos, P. (1996b). Using Partitioning to Speed Up Specific-to-General Rule Induction. In: Proceedings of the AAAI 1996 Workshop on Integrating Multiple Learned Models, pp. 29–34. AAAI Press, Portland (1996).

49. Duda, R. O. and Hart P. E. (1973). Pattern Classification and Scene Analysis. New York: John Wiley & Sons.

50. Eineborg, M. and Boström, H. (2001). Classifying uncovered examples by rule stretching.

    In ILP '01: Proceedings of the 11th International Conference on Inductive Logic Programming, pages 41–50. Springer-Verlag.

51. Elgibreen H. A., Aksoy M. S. (2013) Rules – Tl: A Simple And Improved Rules Algorithm For Incomplete And Large Data. Journal of Theoretical and Applied Information Technology, pp. 28-40.

52. Executive-Order-13402, 2006. Executive Order 13402. http://www.gpo.gov/fdsys/pkg/FR-2006-05-15/pdf/06-4552.pdf [Accessed May 19, 2016].

53. Fawcett, Tom (2006). An Introduction to ROC Analysis. Pattern Recognition Letters. 27 (8): 861–874.

54. Fette I., Sadeh N., Tomasic A. (2007). Learning to detect phishing emails. Proceedings of the 16th international conference on World Wide Web. 649-656.

55. Freund, Y. and Schapire, R.E., (1997) A Decision-Theoretic Generalization of On-Line Learning and an Application to Boosting. Journal of Computer and System Sciences, 55(1), p.119–139.

56. Frank, E., and, Witten, I. (1998) Generating accurate rule sets without global optimisation. Proceedings of the Fifteenth International Conference on Machine Learning, (p. . 144–151). Madison, Wisconsin.

57. Franczak J. M. (2002) Fast Effective Rule Induction Overview. Technical report.

58. Friedman, N., Geiger, D. and Goldszmidt, M. (1997) Bayesian Network Classifiers. Machine Learning - Special issue on learning with probabilistic representations, 29(2-3), pp.131-63.

59. Fürnkranz J., Ullermeier E. H, Menc´ıa E. L., and Brinker K. (2008) Multilabel classi fi cation via calibrated label ranking. *Machine Learning*, 73(2):133–153.

60. Fürnkranz J. (1999) Separate-and-conquer rule learning. Artificial Intelligence Review, 13(1):3–54, 1999.

61. Fürnkranz J. and Widmer G. (1994). Incremental reduced error pruning. In Machine Learning: Proceedings of the Eleventh Annual Conference, New Brunswick, New Jersey,

1994. Morgan Kaufmann.

62. Gaines B. R., Compton P. (1995). Induction of Ripple-Down Rules Applied to Modeling Large Databases. J. Intell. Inf. System. 5(3):211-228.

63. General Assembly of Virginia, 2005. CHAPTER 827. http://leg1.state.va.us/cgi-bin/legp504.exe?051+ful+CHAP0827 [Accessed April 01 2016].

64. Gini, C. (1909). Concentration and dependency ratios (in Italian). English translation in Rivista di Politica Economica, 87 (1997), 769–789.

65. Goldberg, E. (1989) Genetic Algorithms in Search, Optimization, and Machine Learning. MA: Addison Wesley.

66. Google Safe-Browsing, 2010. Google Safe Browsing. http://code.google.com/p/google-safe-browsing/ [Accessed 10 April 2016].

67. Government of Australia (2011) Hackers, Fraudsters and Botnets: Tackling the Problem of Cyber Crime. Report on Inquiry into Cyber Crime, 2011.

68. Gravetter, F. J., & Wallnau, L. B. (2016). Statistics for the behavioral sciences. Cengage Learning.

69. Grossberg (1988) Nonlinear neural networks: Principles, mechanisms, and architectures. Neural Networks, 1(1), p.17–61.

70. Grzymala-Busse J.W. (2010). Rule induction. O. Maimon, L. Rokach (Eds.), Data Mining and Knowledge Discovery Handbook (second ed.), Springer, New York (2010), pp. 249–265.

71. Hall M., Frank E., Holmes G., Pfahringer B., Reutemann P., Witten I. (2009) The WEKA Data Mining Software: An Update; SIGKDD Explorations, Volume 11, Issue 1.

72. Hall M. (1999) Correlation-based Feature Selection for Machine Learning. Thesis, department of computer science, Waikaito University, New Zealand.

73. Hammoud S. (2010). MRSim: A discrete event based MapReduce simulator, Proceedings of the Seventh International Conference on Fuzzy Systems and Knowledge Discovery, (2010) 2993-2997.

74. Hand, D. J., & Yu, K. (2001). Idiot's Bayes—not so stupid after all?. International statistical review, 69(3), 385-398.

75. Hhn J., Hllermeier E., (2009) FURIA: An algorithm for unordered fuzzy rule induction, Data Mining Knowl. Disc., vol. 19, no. 3, pp. 293-319, 2009.

76. Holte, R.C. (1993). Very Simple Classification Rules Perform Well on Most Commonly Used Datasets. Machine Learning, 11, pp 63-90.

77. Information Week (n.d.) http://www.informationweek.com/california-enacts-tough-anti-phishing-law-/d/d-id/1036636?. [Accessed March 17 2016].

78. Jagatic, N., Johnson, A., Jakobsson, M. & Menczer, F.,(2007) Social phishing. Communications of the ACM, 50(10), pp.94-100.

79. Jameel N. Gh., George L. (2013) Detection of Phishing Emails using Feed Forward Neural Network. Journal of Computer Applications 77(7):10-15, September 2013.

80. Jankowski, D., & Jackowski, K. (2014, November). Evolutionary algorithm for decision tree induction. In IFIP International Conference on Computer Information Systems and Industrial Management (pp. 23-32). Springer Berlin Heidelberg.

81. Jevans D. (2003) Anti-Phishing Working Group (APWG): http://www.antiphishing.org/ [Accessed June 20th 2016].

82. Joachims H. (1999) Large-scale support vector machine learning practical, Advances in kernel methods: support vector learning, MIT Press, Cambridge, MA, 1999.

83. Julie S., D., Mandy, H. & Cranor, L.F., 2007. Behavioral Response to Phishing Risk. In The Anti-Phishing Working Groups, 2nd annual eCrime researchers summite, Crime '07. New York, NY, USA, 2007. ACM.

84. Khadi A., Shinde S. (2014) Detection of phishing websites using data mining techniques. International Journal of Engineering Research and Technology, Volume 2(12).

85. Kohavi, Ron (1995). A study of cross-validation and bootstrap for accuracy estimation and model selection. Proceedings of the Fourteenth International Joint Conference on Artificial Intelligence. San Mateo, CA: Morgan Kaufmann. 2 (12): 1137–1143. CiteSeerX 10.1.1.48.529Freely accessible.

86. Kothari R, Dong M (2000) Decision Trees for Classification: A Review and Some New Results. Singapore: World Scientific.

87. Lee J., Bauer L., Mazurek L. M. (2015) The Effectiveness of Security Images in Internet Banking. IEEE Internet Computing (Volume:19, Issue: 1).Pp. 54 – 62.

88. Leung, K. M. (2007). Decision trees and decision rules. Polytechnic University, Department of Computer Science, Finance and Risk Engineering.

89. Li, W., Han, J., and Pei, J. (2001) CMAR: Accurate and efficient classification based on multiple-class association rule. Proceedings of the IEEE International Conference on Data Mining –ICDM, 369-376.

90. Lichman, M. (2013). UCI Machine Learning Repository [http://archive.ics.uci.edu/ml]. Irvine, CA: University of California, School of Information and Computer Science.

91. Liu, H. and Setiono, R. (1995) Chi2: Feature Selection and Discretization of Numeric Attribute. Proceedings of the Seventh IEEE International Conference on Tools with Artificial Intelligence, November 5-8, 1995, pp. 388.

92. Liu, B., Hsu, W., and Ma, Y. (1998) Integrating classification and association rule mining. Proceedings of the Knowledge Discovery and Data Mining Conference-KDD, 80-86. New York.

93. McAfee SiteAdvisor, 2006. McAfee SiteAdvisor. http://www.siteadvisor.com/ [Accessed 19 February 2016].

94. McCall (2007) Gartner, Inc. http://www.gartner.com/newsroom/id/565125 [Accessed June 5th 2016].

95. McFredies, P (n.d.) Phishing. http://www.wordspy.com/words/phishing.asp [Accessed May 15th 2016].

96. Michalski, R.S. (1983). A Theory and Methodology of Inductive Learning, Machine Learning: An Artificial Intelligence Approach, vol. 1, pp. 83-134, 1983.

97. Mohammad R., Thabtah F., McCluskey L., (2015) Tutorial and critical analysis of phishing websites methods. Computer Science Review Journal. Volume 17, August 2015, Pages 1–24 Elsevier.

98. Mohammad R., Thabtah F., McCluskey L., (2014A) Predicting Phishing Websites based on Self-Structuring Neural Network. Journal of Neural Computing and Applications, 25 (2). pp. 443-458. ISSN 0941-0643. Springer.

99. Mohammad R., Thabtah F., McCluskey L., (2014B) Intelligent Rule based Phishing Websites Classification. Journal of Information Security (2), 1-17. ISSN 17518709. IET.

100. Mohammad, R. M., Thabtah, F. & McCluskey, L. (2013)Predicting Phishing Websites using Neural Network trained with Back-Propagation. Las Vigas, World Congress in Computer Science, Computer Engineering, and Applied Computing, pp. 682-686.

101. Mohammad R., Thabtah F., McCluskey L. (2012) An Assessment of Features Related to Phishing Websites using an Automated Technique. In The 7th International Conference for Internet Technology and Secured Transactions (ICITST-2012). pp 492-497., London, UK.

102. Nguyen L. A. T., To B. L., and Nguyen H. K. (2015) An Efficient Approach for Phishing Detection Using Neuro-Fuzzy Model. Journal of Automation and Control Engineering Vol. 3, No. 6, December 2015

103. Pagallo, G. & Haussler, D. (1990). Boolean Feature Discovery in Empirical Learning. Machine Learning 5: 71–99.

104. Pan Y., and Ding X. (2006) Anomaly Based Web Phishing Page Detection. In The 22nd Annual Computer Security Applications Conference (ACSAC). Miami Beach, Florida, USA, 2006. IEEE.

105. Pawlak Z. (1982). Rough Sets. International Journal of Computer and Information Sciences 1982; 11: 341–356.

106. Pawlak Z. Rough Sets (1991). Theoretical Aspects of Reasoning about Data. Kluwer Academic Publishers, Dordrecht, Boston, London, 1991.

107. Pawlak Z., Grzymala-Busse J.W., Slowinski R. and Ziarko, W. (1995) Rough sets. Communications of the ACM 1995; 38: 88– 95.

108. Pazzani, M. & Kibler, D. (1992). The Utility of Knowledge in Inductive Learning. Machine Learning 9: 57–94.

109. Pham D. T. and Soroka A. J. (2007) An Immune-network inspired rule generation algorithm (RULES-IS). In Third Virtual International Conference on Innovative Production Machines and Systems, Whittles Dunbeath, 2007.

110. PhishTank, 2011. PhishTank. http://www.phishtank.com/ [Accessed January 16 2016].

111. Phyu, T. N. (2009) Survey of classification techniques in Data Mining, IMECS 2009 Volume 1 Hong Kong pp. 1-5

112. Pike, G. H. (2006). Lost data: The legal challenges. *Information Today, 23*(10), 1–3.

113. Platt J. (1998) Fast training of SVM using sequential optimization, (Advances in kernel methods – support vector learning, B. Scholkopf, C. Burges, A. Smola eds), MIT Press, Cambridge, 1998, pp. 185-208

114. Powers, D. M. W. (2011). Evaluation: From Precision, Recall and F-Measure to ROC, Informedness, Markedness & Correlation. Journal of Machine Learning Technologies. 2 (1): 37–63.

115. Nahorney, B. (2015) The MessageLabs Intelligence Annual Security Report: 2009 Security Year in Review. http://www.symantec.com/content/en/us/enterprise/other_resources/intelligence-report-06-2015.en-us.pdf [Accessed June 09 2016].

116. Othman O. M. and Bryant C. H. (2015) Pruning Classification Rules with Instance Reduction Methods. International Journal of Machine Learning and Computing, Vol. 5, No. 3, June 2015.

117. Qabajeh I., Thabtah F., Chiclana F. (2015) Dynamic Classification Rules Data Mining Method. Journal of Management Analytics.Volume 2, Issue 3, pp. pages 233-253. Wiley.

118. Quinlan, J. (2002). Data Mining Tools See5 and C5.0. http://www.rulequest.com/see5-info.html.

119. Quinlan, J. (1993) C4.5: Programs for machine learning. San Mateo, CA: Morgan Kaufmann.

120. Quinlan J.R. (1990) Learning Logical Definitions from Relations. Machine Learning, Volume 5, Number 3, 1990.

121. Quinlan J., 1986. Induction of Decision Trees. Mach. Learn. 1, 1 (Mar. 1986), 81-106.

122. Quinlan 1987) J. Ross Quinlan. Simplifying Decision Trees. International Journal of Man-Machine Studies, 27, 221-234, 1987.

123. Quinlan, J. (1979) Discovering rules from large collections of examples: a case study. In Expert Systems in the Micro-electronic Age. Edinburgh, 1979.

124. R Development Core Team (2008) R: A Language and Environment for Statistical Computing. R Foundation for Statistical Computing.

125. Rameshkumar, K., Sambath, M. and Ravi, S. (2013) Relevant association rule mining from medical dataset using new irrelevant rule elimination technique. *Proceedings of ICICES*, 300-304.

126. Ramon J., EDRIessens K., and Croonenborghs T. (2007) Transfer Learning in Reinforcement Learning Problems Through Partial Policy Recycling: Machine Learning. ECML 2007. vol. 4701, J. Kok, J. Koronacki, R. Mantaras, S. Matwin, D. Mladenic, and A. Skowron, Eds., ed: Springer Berlin / Heidelberg, 2007, pp. 699-707.

127. Rader M., Rahman S. (2015) Exploring historical and emerging phishing techniques and mitigating the associated security risks. International Journal of Network Security & Its Applications (IJNSA), Vol.5, No.4, July 2013 DOI : 10.5121/ijnsa.2013.5402 23.

128. Rijnbeek, P. R., & Kors, J. A. (2010). Finding a short and accurate decision rule in disjunctive normal form by exhaustive search. Machine learning, 80(1), 33-62.

129. Ronald, D.J.C., Curtis, C. & Aaron, F.J., (2007) Phishing for user security awareness. Computers & Security, 26(1), pp.73-80.

130. Retun Path (2016) https://blog.returnpath.com/blacklist-basics-the-top-email-blacklists-you-need-to-know-v2/ [Accessed 22 March 2016].

131. Rocchio J. (1971). Relevance feedback in information retrieval. In G. Salton, editor, The SMART Retrieval System: Experiments in Automatic Document Processing, pages 313–323. Prentice-Hall Inc., 1971.

132. Rumelhart, David E.; Hinton, Geoffrey E.; Williams, Ronald J. (1986). Learning representations by back-propagating errors. Nature 323 (6088): 533–536.

133. Sumner M., Frank E., Hall M. (2005) Speeding up logistic model tree induction Knowl. Discov. Databases: Pkdd, 2005 (3721) (2005), pp. 675–683.

134. Sanglerdsinlapachai N., and Rungsawang, A (2010). Using Domain Top-page Similarity Feature in Machine Learning-based Web. In Third International Conference on Knowledge Discovery and Data Mining., 2010. IEEE.

135. SCI2S group, Soft Computing and Intelligent Information Systems, University of Granada, http://sci2s.ugr.es/sicidm [Accessed May 25th 2017]

136. Sheng S., Holbrook M., Arachchilage NAG., Cranor L. Downs J. (2010) Who falls for phish?: a demographic analysis of phishing susceptibility and effectiveness of interventions. In CHI '10 Proceedings of the 28th international conference on Human factors in computing systems. New York, NY, USA, 2010. ACM.

137. Stahl, F., Bramer, M.A. (2014) Random Prism: an alternative to Random Forests. Research and Development in Intelligent Systems XXVIII, 2011, pp 5-18, Springer.

138. Stahl F., Bramer M. (2012) Computationally efficient induction of classification ruleswith the PMCRI and J-PMCRI frameworks, Knowledge-Based Systems 35, (2012) 49–63.

139. Stahl F., and Bramer M. (2008) P-Prism: A Computationally Efficient Approach to Scaling up Classification Rule Induction. Artificial Intelligence in Theory and Practice II, IFIP – The International Federation for Information Processing Volume 276, 2008, pp 77-86.

140. Shawe-Taylor J., Sun S., (2011) A review of optimization methodologies in support vector machines, Neurocomputing 74 (17) (2011) 3609–3618.

141. Taiwiah C. A., and V. Sheng (2013) A study on Multi-label Classification. Advances in Data Mining. Applications and Theoretical Aspects. Lecture Notes in Computer Science, Volume 7987, 137-150.

142. Tan PN., Steinbach M., Kumar V. (2005). Introduction to Data Mining. Published by Addison Wesley.

143. Taylor R.C., (2008). An overview of the Hadoop/MapReduce/HBase framework and its current applications in bioinformatics. BMC Bioinformatics, 11 (2010) S1.

144. Thabtah F., Mohammad R., McCluskey L. (2016A) A Dynamic Self-Structuring Neural Network Model to Combat Phishing. In the Proceedings of the 2016 IEEE World Congress on Computational Intelligence. Vancover, Canada.

145. Thabtah F., Qabajeh I.., Chiclana F. (2016B.) Constrained dynamic rule induction learning. Expert Systems with Applications 63, 74-85.

146. Thabtah F., Hammoud S. Abdeljaber H. (2015) Parallel Associative Classification Data Mining Frameworks Based Mapreduce. To Appear in *Journal of Parallel Processing Letter*. March 2015. World Scientific.

147. Thabtah F., Hadi W., Abdelhamid N., Issa A. (2011) Prediction Phase in Associative Classification. Journal of Knowledge Engineering and Software Engineering. Volume: 21, Issue: 6(2011) pp. 855-876. WorldScinet.

148. Thabtah, F., Cowling, P., and Peng, Y. (2005). MCAR: Multi-class classification based on association rule approach. Proceedings of the 3rd IEEE International Conference on Computer Systems and Applications, 1-7.

149. Thabtah F., Cowling P., and Peng Y. (2006): Multiple Label Classification Rules Approach. Journal of Knowledge and Information System. Volume 9:109-129. Springer.

150. Thabtah, F., Cowling, P., and Peng, Y. (2005) MCAR: Multi-class classification based on association rule approach. Proceedings of the 3rd IEEE International Conference on Computer Systems and Applications, 1-7

151. Thabtah, F., Cowling, P., and Peng, Y. (2004) MMAC: A new multi-class, multi-label associative classification approach. *Proceedings of the Fourth IEEE International Conference on Data Mining (ICDM '04)*, 217-224.

152. Tsoumakas G., Katakis I., and Vlahavas I. (2010) Mining multi-label data. In O. Maimon and L. Rokach, editors, Data Mining and Knowledge Discovery Handbook. Springer, Berlin.

153. Veloso A., Meira W., Gonçalves M., Zaki. M (2007) Multi-label Lazy Associative Classification. Proceedings of the Principles of Data Mining and Knowledge Discovery, PKDD, 605-612.

154. Webb, G. I (1995). OPUS: An Efficient Admissible Algorithm For Unordered SearchOpens in a new window.Journal of Artificial Intelligence Research, 3, 431-465, 1995.

155. Witten I. H. and Frank E. (2005). Data Mining: Practical Machine Learning Tools and Techniques.

156. WOT (2006) Web of Trust. http://www.mywot.com/ [Accessed 24 March 2016].

157. Zaki, M., and Gouda, K. (2003) Fast vertical mining using diffsets. Proceedings of the ninth ACM SIGKDD International Conference on Knowledge Discovery and Data Mining, 326 – 335.

158. Zaki M., Hsiao CJ (2002) CHARM: an efficient algorithm for closed itemset mining. Proceedings of the 2002 Siam International conference on data mining (SDM'02),457–473.

159. Zaremotlagh, S., & Hezarkhani, A. (2017). The use of decision tree induction and artificial neural networks for recognizing the geochemical distribution patterns of LREE in the Choghart deposit, Central Iran. Journal of African Earth Sciences, 128, 37-46.

160. Zhang, M. L., & Zhou, Z. H. (2005, July). A k-nearest neighbor based algorithm for multi-label classification. In Granular Computing, 2005 IEEE International Conference on (Vol. 2, pp. 718-721). IEEE.

INDEX

www.ingramcontent.com/pod-product-compliance
Ingram Content Group UK Ltd.
Pitfield, Milton Keynes, MK11 3LW, UK
UKHW020048200726
13845UKWH00045B/58

9 789349 964945